AF771319

Formation Evaluation with Pre-Digital Well Logs

Formation Evaluation with Pre-Digital Well Logs

RICHARD M. BATEMAN

ELSEVIER

Elsevier
Radarweg 29, PO Box 211, 1000 AE Amsterdam, Netherlands
The Boulevard, Langford Lane, Kidlington, Oxford OX5 1GB, United Kingdom
50 Hampshire Street, 5th Floor, Cambridge, MA 02139, United States

British Library Cataloguing-in-Publication Data
A catalogue record for this book is available from the British Library

Library of Congress Cataloging-in-Publication Data
A catalog record for this book is available from the Library of Congress

ISBN: 978-0-12-820232-6

Publisher: Candice Janco
Acquisitions Editor: Amy Shapiro
Editorial Project Manager: Emerald Li
Production Project Manager: Swapna Srinivasan
Cover Designer: Matthew Limbert

Typeset by MPS Limited, Chennai, India

Contents

Section IV Saturation determination and Archie alternatives 137

Preface

At the time of writing this work, the science and art of subsurface formation evaluation by means of well logging was entering its ninth decade. In 1927 when the first well log was recorded by the Schlumberger brothers, there were no transistors, no digital computers, and no cell phones. At that time, the measurement of subsurface formation properties was limited to a pen and ink trace on a moving roll of paper running in tandem with the progress of a rudimentary formation resistivity measuring device that was hauled up the borehole by a winch. As the art and science of well logging progressed through the 1930s, 1940s, and 1950s, more and more formation properties were measured, and the recording and analysis techniques were improved by leaps and bounds. The decades of the 1960s and 1970s saw the evolution of wireline logging from analog to digital as chronicled in the paper I presented at the 50th SPWLA Symposium held in The Woodlands in 2009[1]. Today the majority of vital petrophysical rock properties are routinely measured, more or less directly, by a combination of sophisticated sensors and advanced modern signal processing and interpretation techniques. However, the same fundamental questions are asked today as they were nine decades ago. The analyst still wants to know porosity, permeability, hydrocarbon type, and saturation in order to reap economic benefits from the subsurface. In general, today, porosity, permeability, and saturation are routinely available quasi-directly from a dazzling array of sensors employing physics, chemistry, biochemistry, acoustics, and sophisticated electronics only dreamed of by the pioneers of yesteryear.

Despite all this, there remain a very large number of pre-digital well logs in the archives of today's oil and gas exploration and production companies. Many of these logs now reside in boxes or in old file cabinets, acquired when assets changed hands in mergers and acquisitions. These records unfortunately do not come with "user instructions," and so they are either ignored or, worse still, taken down to the local dump. Primarily, these old records can provide today's analyst with all the inputs needed for advanced formation evaluation once there is an understanding of the whys and wherefores of these "old E-logs." That is the aim of this

[1] Petrophysical Data Acquisition, Transmission, Recording and Processing: A Brief History of Change from Dots to Digits; SPWLA 50th Annual Logging Symposium held in The Woodlands, Texas, United States, June 21–24, 2009.

work. People who have worked with the pre-digital logs and understand their quirks and foibles are few and growing fewer every day. It is thus with some urgency that I have set about committing to paper my somewhat limited understanding of this lost art.

Richard M. Bateman
November 2019

Acknowledgements

In preparing this collection of "rules of thumb" to be applied to "old E-logs" I have leaned on the wisdom and experience of many, more knowledgeable in these matters than myself. In particular, I would like to thank George Asquith, Roberto Peveraro, Alan Gilchrist, Roland Chemali, Les Bond, Henry Edmundson, Dan Krygowski, Robert Hunt-Grubbe, E.C. Thomas, and many other erstwhile partners in things related to the recording and interpretation of well logs; without their help and encouragement, this work would not have been possible.

SECTION I

The mechanics of reading old E-Log paper prints

CHAPTER 1

Introduction

Abstract

This chapter sets out the purpose of this work and provides a guide to the way the material has been structured. It is not a requirement that the reader necessarily progress through the different chapters in chronological order. In all, there are 18 chapters that have been grouped into seven sections as documented in this first chapter under the paragraph designated "Organization."

Early wireline logging operations. Reproduced from "The Schlumberger Adventure" by Anne Gruner Schlumberger. Published by Arco Publishing, Inc., New York, 1982. ISBN 0-668-05644-4 (pp. 60-61).

1.1 Purpose

This work is intended as a practical "vade mecum" for the working energy professional faced with the problem of making profitable use of well logs recorded as paper prints (analog) in the pre-digital era from the 1930s to the 1970s. Many such well logs are today archived in back room

Formation Evaluation with Pre-Digital Well Logs
DOI: https://doi.org/10.1016/B978-0-12-820232-6.00001-7

filing cabinets and/or on microfiche. The lost art of extracting vital petrophysical "answers" from single point readings taken off wavy curved lines on a paper print can be quite daunting to those schooled in modern digital well-log recordings and continuous data processing. The purpose here, then, is to equip the working geoscientist with the necessary skills to determine the type of device used to record any "legacy" materials available, understand the units of measurement, and "translate" such data into terms that can be used as input to modern log analysis and formation evaluation methods.

This is not yet another book on log analysis and formation evaluation. It is assumed that the reader already has some familiarity with log analysis and petrophysics having completed an undergraduate, or industry, course covering the subject. If that is not the case, then the reader is encouraged to obtain a copy of my exhaustive textbook on the subject.[1]

Along the way, the reader will be introduced to the pioneering thinking of the original "log analysts" who were faced with very limited tools for the job in hand and yet came up with some remarkable methods of "making do" with very limited raw measurements. They were able to arrive at many very useful and practical results on which current analysis methods are based. In many cases, these pioneers found subtle ways of eliminating unknown quantities (e.g., by taking ratios) and were able to provide perfectly usable numbers for porosity and fluid saturations without the benefits of multiple porosity devices, spectroscopy, high-speed data processing or nuclear magnetic resonance.

It is the aim, therefore, to equip the reader with the skills needed to pluck paper prints, recorded 50 plus years ago, from the archives and identify the logging tools used to make the recordings, the units in which they were recorded and then use the data to calculate porosity, permeability, saturation, hydrocarbon type, and lithology.

1.2 Road map

The simple "road map" shown below will help readers to get their bearings. Log analysis objectives for the workers in the analog era were no

[1] Bateman, Richard M., 2012: Openhole Log Analysis and Formation Evaluation – Second Edition, SPE, ISBN 978-1-61399-156-5

different from what they are today. The basic "answers" required included: porosity, permeability, water saturation, hydrocarbon type, and rock type.

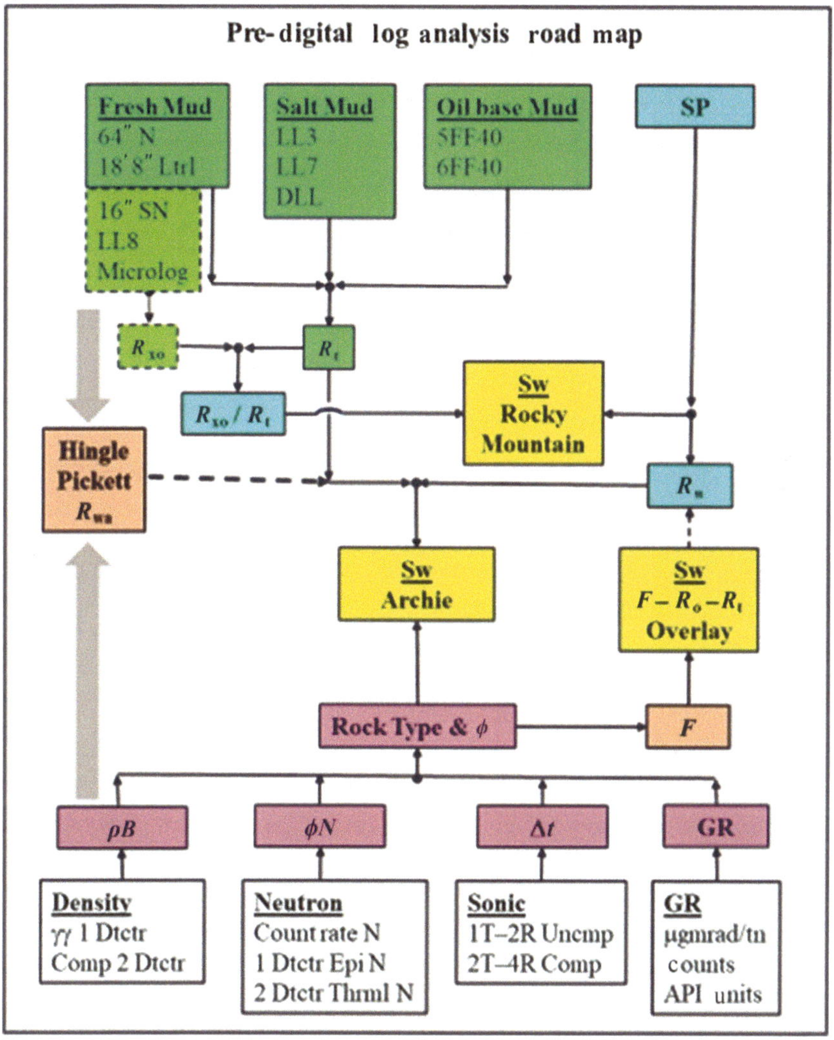

The tools available to find these "answers" were few. The "road map" is arranged so that the historic *resistivity* tools are shown at the top and include those adapted for fresh mud drilling fluids, salt muds, and oil base muds. Their counterparts for *porosity* are shown at the bottom and include

Table 1.1 Basic log measurements analysis methods.

		Porosity	Permeability	Water Saturation	Hydrocarbon Type	Rock Type
Analog Measurements	SP	X	X			X
	Gamma Ray		X			X
	Microlog	X	X			
	Resistivity		X	X		X
	γγ Density	X		X	X	X
	Count Rate Neutron	X		X	X	X
	Sonic	X		X		X
Analysis Methods	Archie			X		
	Rocky Mountain			X		
	F Overlay	X		X		
	Hingle	X		X		
	Pickett			X		X

the early single detector (*abbreviated Dtctr*) neutron and density tools, the single transmitter two receiver sonic tools and the gamma ray tools scaled in strange units, which were precursors to the now familiar GR API units. In the middle, where they meet, several methods are outlined for determining saturations. Table 1.1 summarizes the basic log measurements together with some surprising analysis methods that were used then—and can be used again today—to extract useful data from these early logging measurements.

1.3 Organization

The materials in this work are divided into seven sections. There is no particular order given for the different sections, and they may be used much as a menu at a restaurant. The reader is free to pick the topic of most interest to the task at hand. The seven categories are as follows:

I. The practical "mechanics" of reading old E-log paper prints (Chapters 1 and 2)
II. Sources for the all-important values of R_t and R_{xo} (Chapter 3)
III. Sources for porosity (Chapters 4–9)
IV. Saturation determination (Chapters 10–13)
V. Permeability estimates (Chapter 14)
VI. Lithology determination (Chapter 15)
VII. Miscellaneous (Chapters 16–18)

1.4 History

Up until the 1970s, all well-log recordings were made using analog systems. The logs were made on paper with an ink pen or with a light beam on photographic film. All the recordings were made in real time. There were no recordings made for later playback. In general, all recordings used linear scales. Only very late in the analog era did logarithmic function formers appear to allow resistivity logs to be presented in logarithmic format. Tricks were employed to cover the wide dynamic ranges that resistivity logs could exhibit. “Backup” traces were used that represented the resistivity data at multiples of 5, 10, or 100 times. Log headings were not always as clearly labeled as one might like and in many cases log headers, with the vital scaling information, could become detached from the main body of the paper print log. In such cases, detective work was required to figure out the exact value of formation resistivity.

Table 1.2 lists the chronology of the introduction of logging tool measurements starting in 1927 with the first electric log. Although real-time log analysis was not possible, some limited analysis was possible using compatibly scaled analog curve overlays, which could identify hydrocarbon bearing intervals without the need to use a calculator. For an

Table 1.2 Logging tool history.

1927 First Electric Log	1952 Microlaterolog
1931 First SP Log	1950 Laterolog
1931 Sidewall Cores	1956 γγ-Density (Uncompensated)
1932 Bullet Perforating	1957 Sonic (uncompenstaed)
1933 Temperature Log	1962 Sidewall Neutron
1936 SP Dipmeter	1962 Compensated Density
1938 First Gamma Ray Log	1963 Dual Induction
1940 Count-rate Neutron	1964 Pulsed Neutron
1941 Archie's Law Published	1966 Compensated Neutron
1947 Resistivity Dipmeter	1969 Spectral Gamma Ray
1947 Induction	1972 Dual Laterolog
1948 Microlog	1977 First Digital Logging Truck
1949 Laterolog	

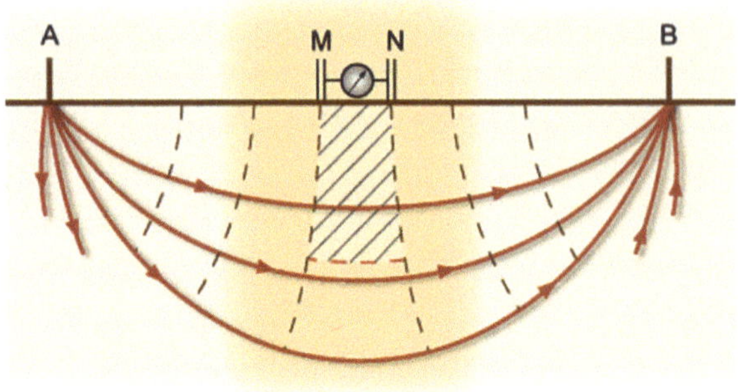

Figure 1.1 Surface prospecting apparatus. Image courtesy Schlumberger.

exhausting history of the evolution of well logging, the reader is referred to my paper given at the 50th Annual SPWLA Symposium.[2]

Many authors have chronicled the pioneering work of Conrad and Marcel Schlumberger[3] who started out by performing surface electrical surveys to profile subsurface formation shape and content. Fig. 1.1 shows a schematic of the setup consisting of four electrodes that were driven into the ground, two current electrodes, A and B, and two potential electrodes M and N.

The surface prospecting process required the mechanical "planting" of the array sequentially along the line to be surveyed as shown in Fig. 1.2.

The key advance from surface electrical prospecting to well logging came from the simple act of rotating the four-electrode system through 90 degrees as shown in Fig. 1.3.

The first electric log was recorded by Henri Doll, Roger Jost, and Charles Scheibli in Pechelbronn in 1927. They took stationary (point) measurements every meter of the wellbore recording a total of 150 measurements in a period of 3½ hours. (Fig. 1.4)

1.5 Analysis methodology

Log analysis in clean formations follows the familiar Archie model as illustrated in Fig. 1.5, where porosity is defined by one or more of the traditional "porosity" devices, and the water-filled pore space is quantified by

[2] Bateman, Richard M., 2009, Petrophysical Data Acquisition, Transmission, Recording and Processing: A Brief History of Change from Dots to Digits, SPWLA 50th Annual Symposium held at The Woodlands, June 21-24, 2009.

[3] For example, Oristaglio and Dorozynski or Allaud and Martin

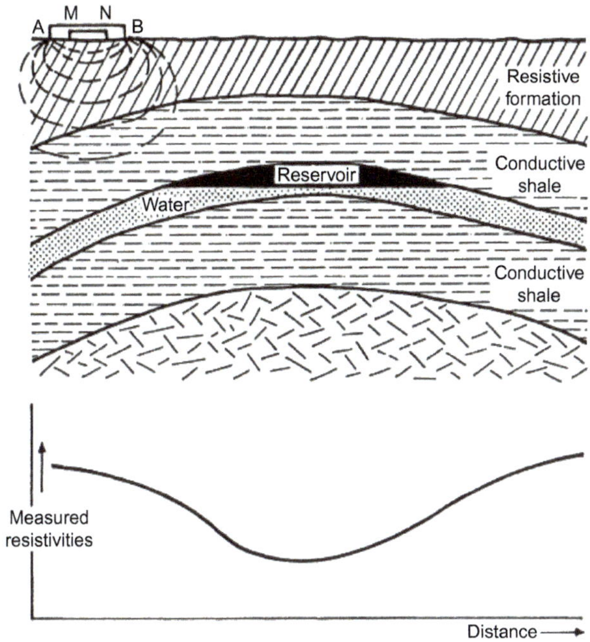

Figure 1.2 Example of surface prospecting defining an anticline. Image courtesy Schlumberger.

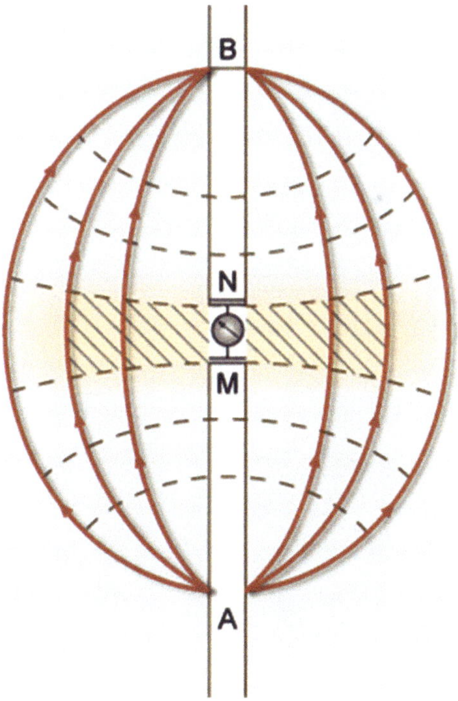

Figure 1.3 Four-electrode system for well logging. Image courtesy Schlumberger.

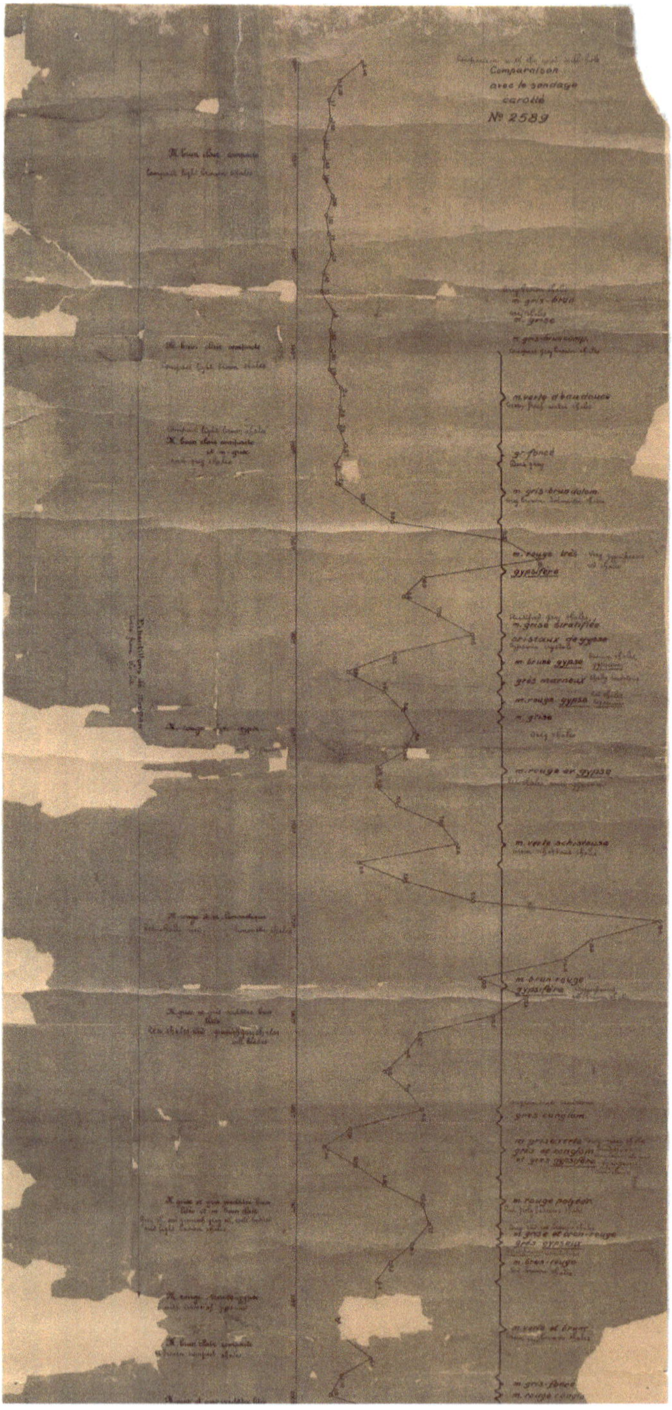

Figure 1.4 The first electric log. Image courtesy Schlumberger.

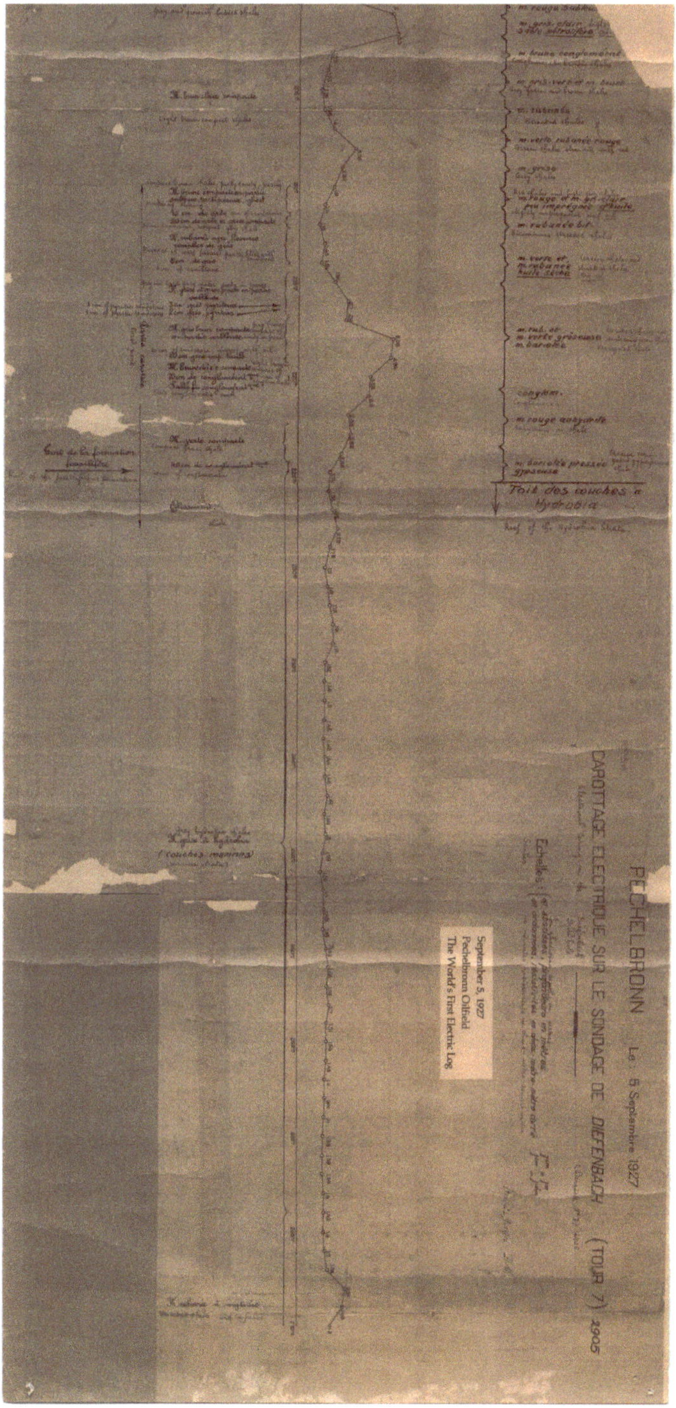

Figure 1.4 (Continued)

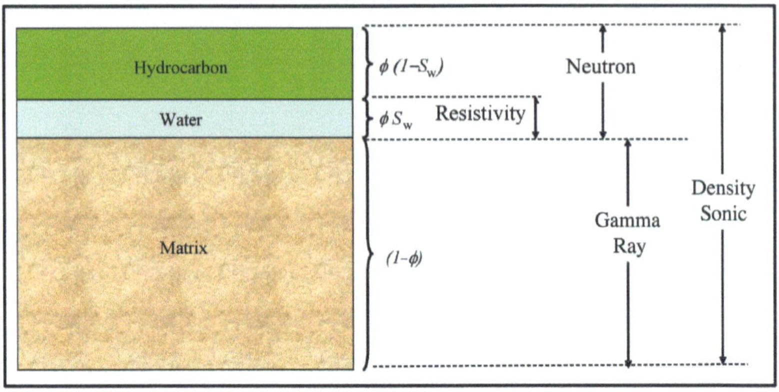

Figure 1.5 Traditional "Archie" clean formation model.

a resistivity device. The conductive path through the porous rock is quantified by the bulk volume of water (BVW) and the rock conductivity, C_t, depends on the square of the BVW and the water conductivity, C_w.

$$C_t = C_w(\phi S_w)^2$$

The same equation in terms of resistivity of the familiar Archie equation:

$$S_w = \sqrt[2]{\frac{aR_w}{\phi^m R_t}}$$

1.6 Reading analog log prints

Much of the difficulty of dealing with old analog prints arises from the scaling for the curves the analyst is attempting to read. Thus a large part of the process of making any kind log analysis is in fact wrapped up in things other than log analysis per se. More than half the battle is determining what kind of a logging device was used, what units of measurement were used, and what is the scale for each curve. In the chapters that follow, each commonly encountered old log will be discussed in detail. Common, to all but the very oldest log prints, is the standardized API log grid. This is illustrated in Fig. 1.6 (not to scale). It consists of three "tracks," each 2½ in. wide with a "depth track," half an inch wide, between the first (left-hand) track and the other two, conventionally known at "Tracks 2 and 3."

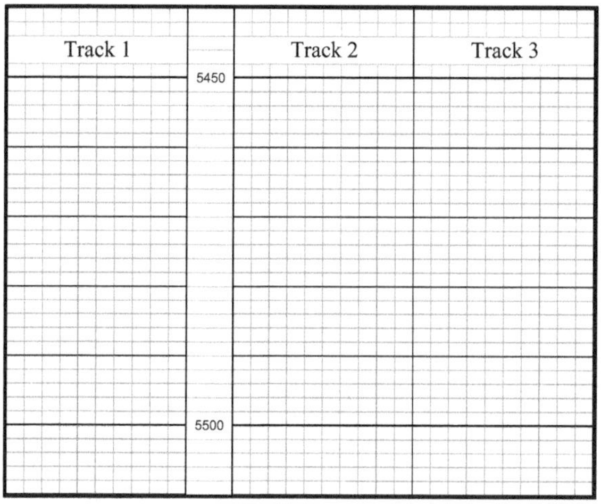

Figure 1.6 API log grid.

1.7 Depth coding

Depth coding on analog prints depended both on the driller's depth units (English or Metric) and on the depth scaling required which, traditionally, was either 1:240 (equivalent to 5 in. of paper to 100 ft of wellbore) or 1:600 (2″/100′) or even 1:1200 (1″/100′).

Fig. 1.7 (a) shows an example log grid on a 5″ scaling where there are light depth lines every 2 ft, heavier depth lines every 10 ft. Fig. 1.7 (b) shows the equivalent depth line coding for metric logs where the depth scaling is usually 1:200 for detailed analysis and either 1:500 or 1:1000 for a condensed "big picture" look over an entire logged section. Note that on the metric log prints, the depth lines are every 1 m apart with bolder lines every 5 m and even bolder ones at every 25 m.

1.8 Resistivity scaling

Before the introduction of the induction log (1956) resistivity scaling for E-logs, micrologs, and laterologs was all linear but with specially scaled "backup" traces. Today we seldom read the value of any log parameter directly from the log itself. We pull up the digital file, go to the depth of interest and read, for example, that the deep induction recorded 15.5 Ωm. At worst we may actually look at the induction log paper print

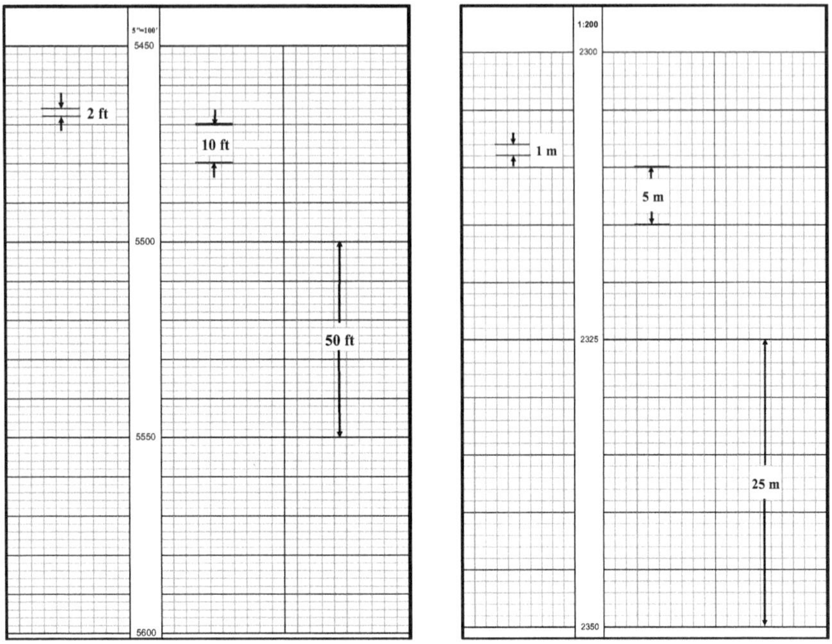

Figure 1.7 (a) Depth line coding for 5″ log and (b) metric log.

(or the log display on the computer screen) and read the value understanding that the scaling is logarithmic from say 0.2 to 2000 across two log tracks. However, life was not so simple when logarithmic scaling was unavailable.

To accommodate a wide dynamic range of possible resistivity values over a logged interval, the early electric logs employed two "tricks." If a regular scale was selected to record resistivities from 0 to 20 Ωm across a given track, then two other traces might also appear. If the formation resistivity rose higher than 20, then the main curve would disappear off the high side (usually the right-hand side) of the track in question and a "backup" trace would appear on a scaling of 0–200, that is, 10 times less sensitive. It should be noted that while the main resistivity curve was within the limits of 0–20, the backup would remain hidden and would appear only when the resistivity exceeded the higher limit of the main curve, in this case the 20 Ωm. At the other end of the scale, in very low resistivity formations, a third trace could appear on an amplified scale of 0–5 Ωm. This too would only make its ephemeral appearance when the formation resistivity was within the limits of 0 and 5 Ω,

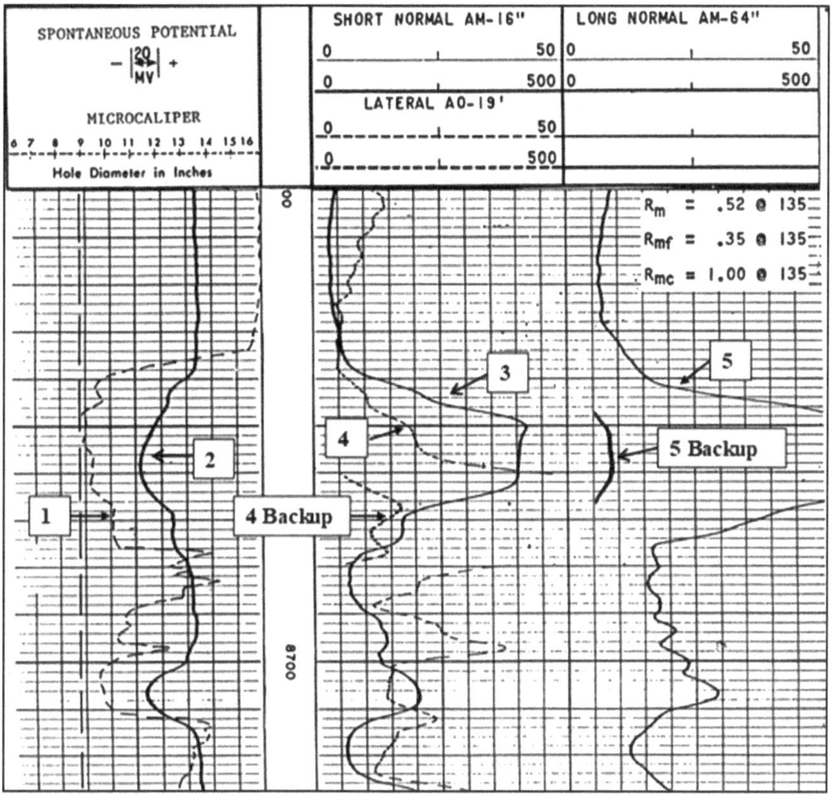

Figure 1.8 Old E-log resistivity scaling example.

for example. These conventions are more easily understood with reference to Fig. 1.8, which shows a short section of an electric log on which five measurements are visible. From the left to right:

1. Dashed caliper (6″–16″)
2. Solid SP (20 mV/chart division)
3. Solid 16″ short normal (0–50 Ωm)
4. Dashed 19′ lateral[4] (0–50 Ωm) and its backup trace (0–500 Ωm)
5. Solid 64″ long normal (0–50 Ωm) and its backup trace (0–500 Ωm)

Pre-digital logs were limited to identifying curve traces by simple coding: solid, dotted, dashed, light, or bold. The same functions today are performed by color as an additional identifier. However, color printers in logging units came much later.

[4] Strictly an 18′8″ lateral, but in this case, it is labeled on the log as just 19′.

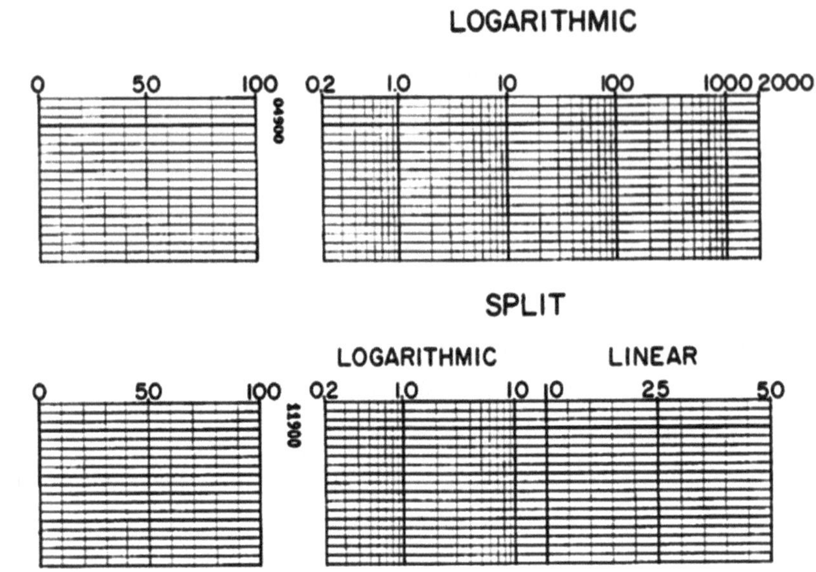

Figure 1.9 Induction log "four-decade" and "split grid" resistivity scaling.

A special note is called for on the matter of the "backup" curves. On Fig. 1.8, it is noticeable that in the depth interval 8648–8666, the long normal backup only appears on the log after the main curve has disappeared off scale to the right at its maximum value of 50 Ωm. The backup curve appears starting at the same ohmic value (50 Ωm) which, on the backup scale of 0–500 starts one chart division in from the left-hand side of the track.

Logarithmic scaling removed the need for backup traces and come in two varieties, as shown in Fig. 1.9.

Where the dynamic range of resistivities over a given logged interval is relatively small, as might be the case with high porosity shaly sand sequences, then the "split grid" scaled from 0.2–20 Ωm is used. Where the dynamic range is larger, as may be the case in tight carbonate reservoirs, then the "four-decade grid" is more common, and the resistivity scaling is from 0.2 to 2000 Ωm.

1.9 Genealogy of wireline service companies

During the pre-digital age, there were no less than 10 different wireline service companies whose logs may well surface in the file cabinets you

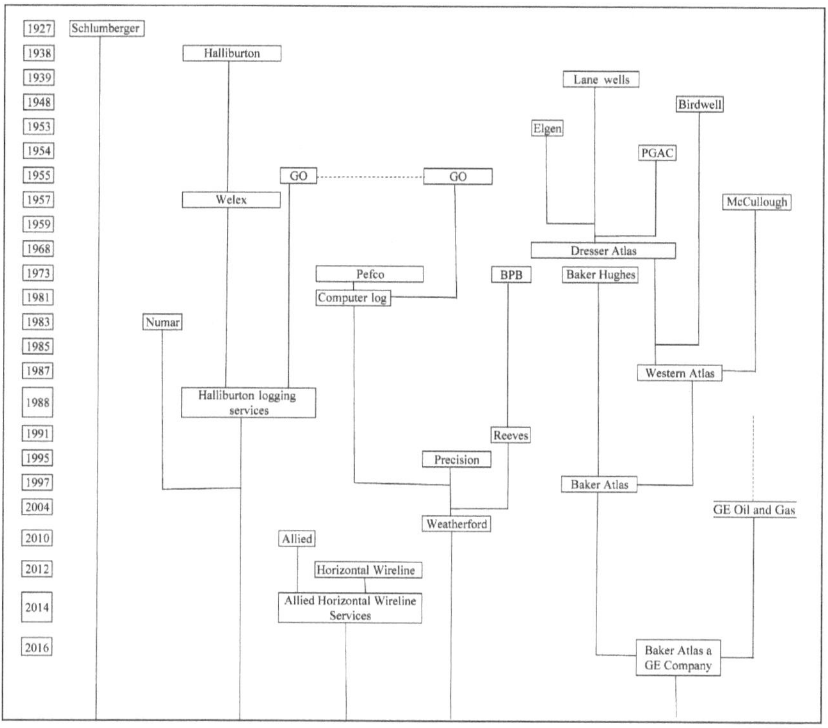

Figure 1.10 The genealogy of wireline service companies. Based on original work by Richard Bateman and Daniel Krygowksi of The Discovery Group, Denver, Colorado.

may be working with. Of historical interest is the chart, shown as Fig. 1.10, that gives the genealogy of these fine companies and also maps the sequence of mergers, acquisitions, and name changes that have occurred in this group since the late 1920s. It also extends the continuing changes so that today's reader may see that the logs being worked with today have roots that may be long forgotten.

Further reading

Barnes, K.B., 1949. New logging method. Oil Gas J. 48 (34), 35.

Bateman, R.M., 2009. Petrophysical data acquisition, transmission, recording and processing: a brief history of change from dots to digits. SPWLA 50th Annual Symposium Held at The Woodlands. 21–24 June.

Brown, F.H., U.S. Patent 274,882.

Dale, C.R., U.S. Patent 2,203,720.

Deegan, C.J., 1950. New logging method. Oil Gas J. 48 (43), 23.

Doll, H.G., Martin, M., Tixier, M.P., 1959. Review of the progress of well logging since the fourth world petroleum congress. Sec. 1, Paper 35, Proc. of Fifth World Petroleum Congress. New York City, pp. 645–661.
Fox, R.W., 1830. On the electromagnetic properties of metalliferous veins in the mines of cornwall. Trans. Royal.—SOC. London 25.
Houston Chapter Society of Professional Well Log Analysts, 1979. The Art of Ancient Log Analysis. Society of Professional Well Log Analysts.
Schlumberger, C., French Patent 450,784.

CHAPTER 2

Applications

Abstract

Here, common elements of log analysis and formation evaluation using pre-digital well logs are discussed. Today, many of the steps, the analyst needs to take for an effective analysis are invisible—but that does not mean that when using the older logs they can be ignored. These well-chosen remarks should ensure that the explorer of the past gets off on a good footing.

Early Logging Equipment with Continuous Recording—1930. Image courtesy Schlumberger.

2.1 Workflow

When faced with a drawer full of analog prints, an ordered approach to a coherent formation evaluation plan is useful, if not essential. The modern log analyst needs to understand that there are many differences between a digital log dataset, that is the day-to-day norm now, and what was available in the pre-digital era. The principal differences are listed in Table 2.1.

To analyze a pre-digital dataset successfully, the modern analyst must therefore adjust to a different modus operandi. A good way to start would

Formation Evaluation with Pre-Digital Well Logs
DOI: https://doi.org/10.1016/B978-0-12-820232-6.00002-9

Table 2.1 "Then and now" differences.

Today	Then
Data exists as a digital recording in a computer file.	Analog curves on paper prints.
All logging information is gathered on a single run with a combo tool hence: all logging data mutually on depth.	Each separate logging curve was recorded on a separate run into the hole resulting in inter-curve depth discrepancies.
All log data is in one place - the digital file.	Reconnoitering was required to research the log heading of the primary run to see what "other services" were run in the same hole - and then find the corresponding log prints!
Log data is sampled at a fixed depth increment (e.g.: one sample every 6 ins or 10 cms).	Analog data was continuously recorded as a curve plot versus depth that could be on a fine, or coarse, depth scale (e.g. 2" or 5"/100' or 1/200 or 1/1000 for metric depths).
All common data is in one place (e.g. the well location, mud type and resistivity, temperatures, survey depths and spans etc.).	Heading data on the primary log did not always include data relevant to subsequent, separate, logs run in the same well (e.g. neutron, density and sonic calibration settings).
Log analysis is performed as a continuous "batch" process.	Analysis was performed on a point by point basis by hand using charts and/or nomograms.

thus be to lower expectations and be content with a few well-chosen data points selected on the basis of local knowledge and, perhaps, more recently drilled neighboring wells, which do have modern, digital, log suites. At each of these data points, the analyst should collect sufficient resistivity, porosity, and lithological indicator data to allow a reasonable analysis based on the rock type and water saturation. The relationships used will be the same as those in use today (e.g., Archie's equation), but the inputs to them may require unaccustomed skills and manipulations. It is the aim of this work to provide easy to follow guidance to perform these practical steps by following a few rules of thumb. So equipped the analyst should be able to confidently identify by-passed pay intervals as well as porous and permeable intervals suitable for water flooding, waste water injection or gas storage, etc. With luck and skill organic-rich shales may even be identified from pre-digital logs.

2.2 Methodology

There are several major log analysis topics documented in Table 2.2 to be addressed, which are summarized here and then expanded on in much more detail in the sections that follow.

Table 2.2 Summary of data sources and analysis procedures.

Log Heading	Well location
	Depths (TD, Casing, First and Last log readings)
	Mud related resistivities and temperatures
	Names of other logging services run in same well
Resistivity	E-log spacings: 16" Short Normal $\approx R_{xo}$
	64" Normal $\approx R_i$
	18' 8" Lateral $\approx R_t$
	Laterologs: Laterolog 3 $\approx R_t$
	Laterolog 7 $\approx R_t$
	Laterolog 8 $\approx R_{xo}$
	Micrologs (Permeability & Pay counting)
	Induction logs: 5FF40 and 6FF40
	Dual Induction logs: Rt corrections for invasion
Porosity	Neutron Counts $\rightarrow \phi_N$
	$\gamma\gamma$ Counts $\rightarrow \rho_B \rightarrow \phi$
	$\Delta t \rightarrow \phi_S$
	Microlog $\rightarrow$ F $\rightarrow \phi$
Lithology	$\phi_N / \rho_B / \Delta t$ combinations + SP & GR
Saturation	Archie
	Rocky Mountain Method (R_{xo}/R_t vs SP)
	F Overlay
Rw	Hingle Plot
	Pickett Plot
	SP
Permeability	SP linked to core
	Timur / Wyllie & Rose (S_w + ϕ)
	Raymer and Freeman
	Resistivity Gradients

In the early days of log interpretation, matters were quite simple. One of the pioneers (Leonardon) published a guide to wireline log interpretation as follows:

Resistivity Log	SP Log	Probable Conclusion
Low	No SP	Shales
Low	Large SP	Saltwater sand
High	Low SP	Freshwater sand
Good	Moderately large SP	Possible oil sand
Very high	No SP	Hard rock; very compact sand with sweet water

This qualitative guide was all that was needed when the only logs available were a single resistivity profile along with an SP. Indeed, there is an apocryphal story of a revered log company log analyst who would report to work religiously every morning, take a key from his pocket and unlock a drawer in his desk, glance quickly into it, and then close it back up under lock and key until the next day. After many years, the analyst retired, and, in his goodbye dinner, his colleagues pleaded with him to reveal the secret he had in his drawer that he guarded under lock and key. After much cajoling, he finally relented and unlocked he drawer to reveal a 3 by 5 index card on which he had printed:

REMEMBER: SP is on the left—Resistivity is on the right

2.3 The importance of mud

Mud is an essential element in the whole process of measuring formation resistivity with electrical logging tools. Without conductive mud in the drill hole, there would be no way to cause a direct electrical current to flow through the earth formations penetrated by the drill bit, neither would there be a way for monitoring electrodes to detect subtle potential differences in the borehole caused by such current flows. Indeed, the need for induction logging was precisely because direct electrical current will not flow in holes drilled with nonconductive oil base muds. The same can be said for air-drilled or "empty" holes. Thus the electrical properties of drilling muds are an important key to the interpretation of electrical logs.

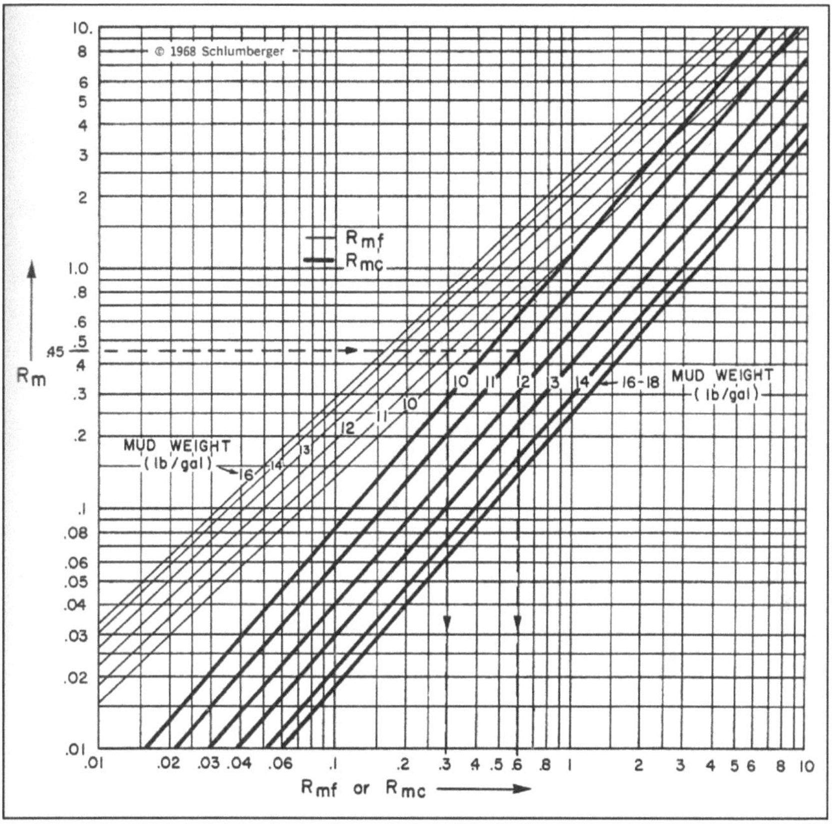

Figure 2.1 Mud, mud cake, and mud filtrate resistivity as function of mud weight. Image Courtesy Schlumberger.

The mud properties that the analyst needs to perform a meaningful interpretation include mud weight and the resistivities of the mud, R_m, the mud cake, R_{mc}, and the mud filtrate, R_{mf}, together with their respective measure temperatures. Under normal circumstances, these measurements are made at the time of logging and noted on the log header. However, in the case of legacy materials, the mud-related data may be incomplete. In such cases, it is useful to have a fallback method to approximate the missing data. Fig. 2.1 serves this purpose. It provides a plot with R_m in the Y-axis and R_{mf} and R_{mc} in the X-axis for a range of mud weights from 10 to 18 lb./gal. The relationships plotted can be fitted to the expressions:

$$R_{mf} = R_m{}^{1.07} x 10^{\left[\frac{(9-W)}{13}\right]}$$

$$R_{mc} = R_m{}^{0.88} x10^{\left[\frac{(W-10.4)}{7.6}\right]}$$

Another, simpler, approximation for NaCl muds gives the mud filtrate and mud cake resistivities in terms of the mud resistivity itself:

$$R_{mf} = 0.75R_m \text{ and } R_{mc} = 1.5R_m$$

Needless to say the proper application of the R_m, R_{mf}, and R_{mc} numbers calls for correction to formation temperature. In this work, it is assumed that the reader is familiar with the mechanics of estimating formation temperatures given surface temperatures and temperature gradients. Likewise, the use of the Arps[1] equation for temperature corrections to fluid resistivities:

$$R_2 = R_1 \frac{(T_1 + 7)}{(T_2 + 7)}$$

These mud-related resistivity parameters appear in a plethora of raw data correction charts and interpretation graphics that will appear along the way as the reader follows this work.

2.4 Invasion, bed thickness, and shoulder beds

The geometries of the beds to be analyzed are important. Electric logging systems have finite spacings of their current electrodes and potential difference sensing electrodes and thus the resistivity value appearing on a log depends not only on the resistivity of the bed of interest but also on many other variables such as:

Hole size and R_m

The resistivity of the beds either side of the bed of interest (shoulder beds), R_s

The invasion diameter, d_i

The resistivity of the invaded zone[2], R_{xo}, and finally the parameter of greatest interest

The resistivity of the uninvaded zone, R_t

[1] This version of the Arps equation is accurate for temperatures in degrees Fahrenheit.

[2] In the early literature, the resistivity of the invaded zone was also referred to as R_i. In this work, the modern usage of R_{xo} will be used.

The basic electric logging tools are unfocused devices and the electric currents follow the path of least resistance, be that in the mud column, through the shoulder beds, through the invaded zone or, where one would hope it flows, through the undisturbed formation, far from the borehole wall.

2.5 Focused resistivity devices

To the early electric loggers, it became obvious that a better reading of formation resistivity could be obtained if the current flow was focused. These devices used a number of "tricks" to ensure that the lion's share of the current leaving the electrodes went into the formation rather than elsewhere. These are described in more detail in Chapter 3 and include:

Laterolog 3
Laterolog 7
Dual Laterolog (much later)
Microlog
Microlaterolog

Although focused, these devices still had to send current through the mud column to reach the formation. The induction devices side stepped this issue by inducing current flow in the formation without the need for anything conductive in the borehole, which could in fact be empty (air drilled) or filled with nonconductive oil base mud. These devices are described in more detail in Chapter 3 and include:

5FF40 Induction
6FF40 Induction
Dual Induction—LL8
Dual Induction—SFL

2.6 Porosity and lithology devices

The very earliest "porosity" device was the SP in as much as it was used as a "binary" indicator of the presence or absence of a porous zone. If there was an SP deflection, it was porous; if there was no SP deflection—it was an impervious shale. This is reflected in Leonardon's table. To some extent, the gamma ray served a similar purpose in that it too was a good discriminator between carbonates, sands, and shales. The mainstays of the pre-digital era for porosity were the Sonic, Neutron, and Density logs, which are covered in more detail in Section III that includes

Chapters 4 through 9 covering Sonic, Count-rate Neutron, Gamma Gamma (uncompensated) Density, Sidewall (Epithermal) Neutron, Compensated (Thermal) Neutron, Microlog, and SP.

A common theme in the evolution of these "porosity" tools is the step from single sources and detectors to dual detectors and, in the case of the Sonic tool, dual transmitters. With each step in the evolution, the end product was a reading of a formation evaluation parameter that was more exact and less troubled by the environmental distortions due to hole size, mud weight and type, and the presence of mud cake in front of the most important zones—the porous and permeable ones.

The same devices that were used for porosity were also used for lithological identification. These rock typing methods are documented in Section VI under Chapter 15 and include the Neutron, Density, and Sonic crossplots as well as gamma ray and SP applications in matters related to rock typing.

2.7 Water saturation and permeability

Water saturation methods with pre-digital logs are covered in Section IV where Chapters 10 through 13 address the Rocky Mountain method, the Pickett and Hingle plots, log analysis in empty holes, and the quick-look methods afforded by the R_{wa} and F-Overlay techniques. Section IV also addresses the parallel task of determining the all-important parameter, R_w.

Permeability is covered in Section V in Chapter 14, which takes the reader through five easily applicable methods that include Timur, Wyllie & Rose, Raymer and Freeman, Resistivity Gradients, and the SP log. All of these are applicable to the measurements and analysis available from pre-digital well logs.

2.8 Quick-look example log analysis

The log shown in Figure 1.8 (see Chapter 1) also serves to show that simple calculations can produce valuable answers. A marked up copy of the log is reproduced here as Fig. 2.2. As we will see in Chapter 3 (on resistivity logging), the depth of investigation of the various old electric log measurements parallels the actual spacing (be it an AM or an AO or other device). Thus, of the three resistivities available in our current example, the 16″ short normal reads the resistivity of a volume of rock closest to the

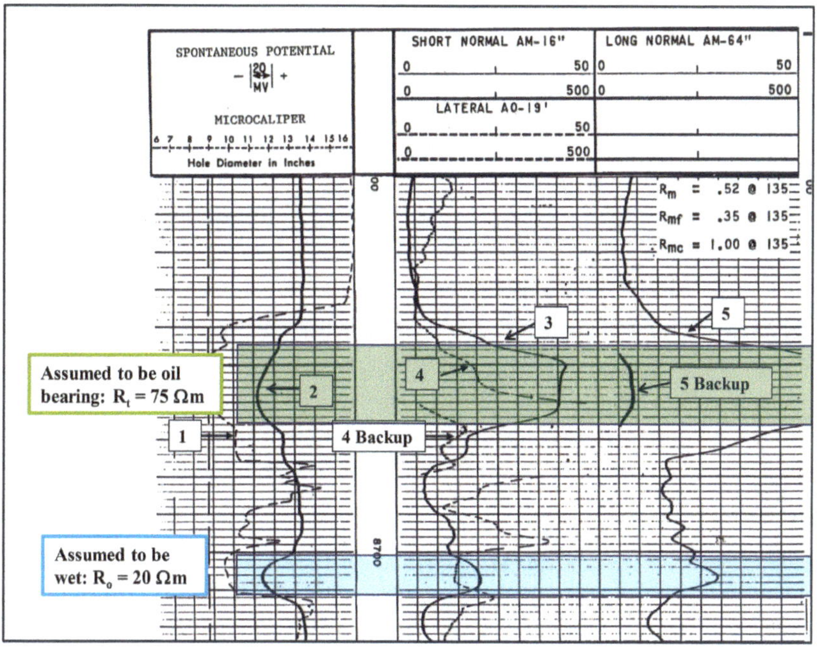

Figure 2.2 Quick-look analysis from electric log.

borehole, the 19′ lateral the resistivity farthest from the borehole and the long normal (64″) somewhere in-between. In general, one could assume that:

$R_{16''SN} \approx R_{xo}$

$R_{64''\ LN} \approx R_i$ (something between R_{xo} and R_t)

$R_{19'\ \mathrm{lateral}} \approx R_t$

By observation, there are two porous and permeable zones. The lower zone (8700–8710) shows low resistivities on all three traces and is probably wet. The upper zone (8648–8666) shows high resistivities and is probably oil bearing.

The mud system is water based, and the curve separations imply that mud filtrate invasion has occurred before logging. If the Archie saturation equation is used in its most basic form a "Quick-look (QL)" value of the water, saturation can be computed by taking the square root of the resistivity of the wet rock to that of the oil-bearing rock, that is, $S_w = (R_o/R_t)^{0.5}$. This assumes that the porosity in both zones is the same—which may not be true.

Zone	Assumed	$R_{16'' SN}$	$R_{64'' LN}$	$R_{19' Lateral}$
Lower	Wet	20	30	15
Upper	Oil Bearing	40	75	75
QL Water Saturation	$\sqrt{(R_{Lower}/R_{upper})}$	71%	63%	45%
Resistivty corresponds to		R_{xo}	transition	R_t

On this basis, and using the 19′ lateral as a measure of R_t, it would appear that the upper zone is oil bearing with a water saturation close to 45%. A similar calculation for the other two resistivity traces confirms a normal invasion profile with an S_{xo} value of 71%.

Further reading

Schlumberger, C., Schlumberger, M., 1929. Electrical logs and correlations in drill holes. Pet. Eng. 86, 87, 90.

SECTION II

Sources for Rt & Rxo

CHAPTER 3

Formation resistivity

Abstract

Of all the pre-digital logs available today, the most commonly found in those back-room file cabinets will be the "old E-logs" or *electric logs*. By far the greatest challenge for the modern log analyst has been to find the correct value of the undisturbed formation resistivity, R_t. It is not an easy task and it is hoped that the treatment of this thorny subject, given here, strikes the right balance between what is practical and useful and yet avoids a descent into the more arcane and tortuous rules of thumb that can be found elsewhere in the literature on this subject.

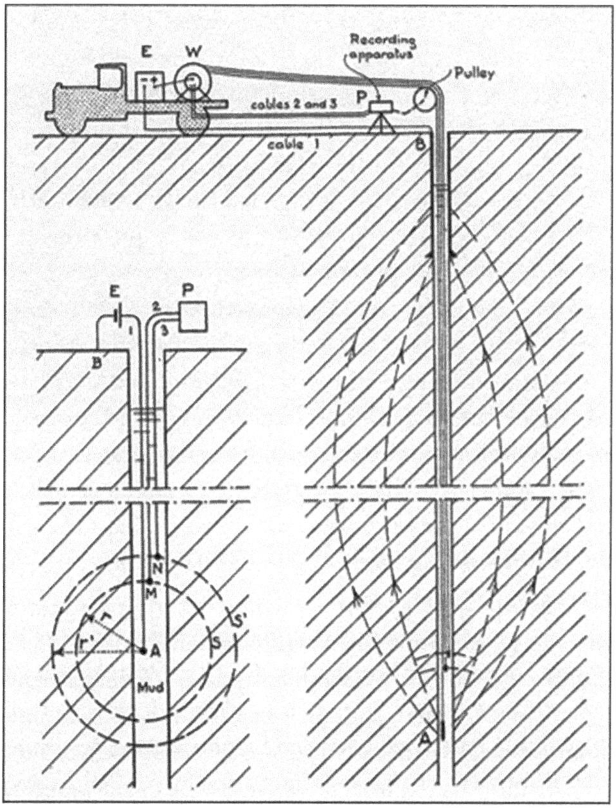

Reprinted with permission of the Society of Petroleum Engineers from Transactions, American Institute of Mining and Metallurgical Engineers, *1934.*

Formation Evaluation with Pre-Digital Well Logs
DOI: https://doi.org/10.1016/B978-0-12-820232-6.00003-0

3.1 Thirty thousand foot view of electric log interpretation

The interpretation of the original, simple, electric logs in many ways is a more difficult task than that of interpreting the multi-component well logging suites we are used to working with today. It is sobering to learn that a full understanding of the intricacies of the recordings of a few simple tools, with no more than four electrodes, could fill a book-size document written in 1950. But such is the case. This landmark publication was issued by Schlumberger Well Surveying Corporation with the title:

Interpretation handbook for resistivity logs
Schlumberger Document Number 4—July 1950

This 148-page *tour de force* describes, in minute detail, every nuance of the interplay of all the factors that influence what is recorded on an electric log and how such recordings may be interpreted to give the analyst robust numbers for bed thicknesses and formation resistivities in a mind boggling variety of borehole and subsurface conditions. If this *Document 4* had been prepared by a university student, it could have been as a PhD thesis. The level of scholarship and detail in this publication is truly impressive. As a scientific treatise it shines. As a practical guide to log interpretation of old e-logs it is unsurpassed.

Unfortunately, lack of space in this current work means that the reader will have to be content with the reproduction, in the appendix to this chapter, of the Introduction, and the Table of Contents of the classic Document 4. Few copies remain in circulation today, but the determined analysts would do well to find one.

3.2 The problems of measuring formation resistivity

To properly asses the undisturbed formation resistivity a number of problems have to be overcome. They include the effects of:

- The borehole itself—its diameter (d_h) and fluid content resistivity (R_m),
- The influence of mud filtrate invasion (with resistivity R_{mf}) radially into the formation to a dimeter (d_i), if porous and permeable, which introduces between the wellbore and the undisturbed formation an annular region with a resistivity R_{xo} that is different from the more

distant, undisturbed, region with the resistivity that is sought, which is R_t.

- The effects of the thickness of the bed of interest (*h*), and
- The effects of the surrounding (or "shoulder") beds with a different resistivity, R_s.

Fig. 3.1 illustrates these conflicting demands on both logging tool designers and log analysts charged with making deductions about the true formation resistivity, R_t.

3.3 Tools available

Over the years "families" of resistivity logging tools have proliferated. Broadly speaking they may be categorized as:

- Unfocussed electric (short normal [SN] and long normal and long laterals), collectively "E -logs"
- Focused electric (guard or laterologs -3, -7, and -8) and dual laterologs (shallow and deep)
- Micro-resistivity (microlog, microlaterolog, proximity log, micro-spherically focused log [SFL])
- Induction (5FF40 and 6FF40) and dual inductions
- Dielectric

In this work, analysis will be restricted to those devices that were in use up until the late 1970s. It will also be assumed that the basics of log analysis are understood by the reader. However, it is worthwhile to make a brief summary here regarding the principles involved and to reignite interest in unlocking the secrets hidden in "old logs."

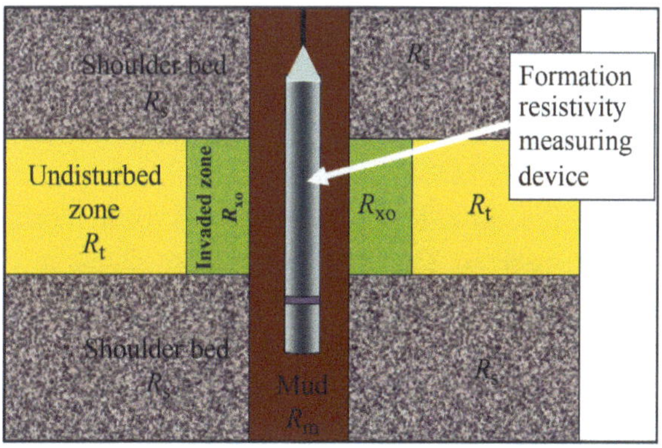

Figure 3.1 Schematic of a generalized formation resistivity measuring device.

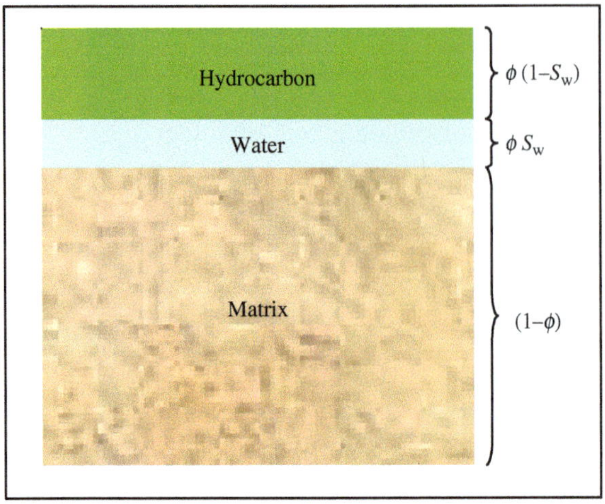

Figure 3.2 Repartition between rock, hydrocarbon, and water.

3.4 Formation resistivity logging recap

Resistivity is measured in Ohm meter^2/meter ($\Omega m^2/m$). It is the voltage required to cause one amp to pass through a cube one meter long and one meter square. Typical formation resistivities lie between 0.5 and 1000 $\Omega m^2/m$. Resistivities in shaly sands typically lie between 0.5 and 50 $\Omega m^2/m$, whereas carbonates are more resistive—between 100 and 1000 $\Omega m^2/m$. Evaporites, being very good electrical insulators, may log in at several thousand $\Omega m^2/m$. Formation water resistivities range from 0.01 to 10 $\Omega m^2/m$ (brine → fresh water) and sea water clocks in at 0.35 $\Omega m^2/m$ at 75°F.[1]

Electrical resistivity is a sensitive parameter, useful for distinguishing between salt water and hydrocarbons since salt water is a fair conductor but both oil and gas are insulators and (most) rock matrices are also nonconductors. A simplified schematic shows the volumetric repartition in Fig. 3.2.

The key concepts include the physical dimensions of the sundry zones, their resistivities (as rock/fluid mixtures), the fluid content type, and the saturations (if more than one fluid is present), not to mention the resistivities of the fluids themselves. This information is available in diagrammatic form in Fig. 3.3. A simplified way to account for all of this is given in Table 3.1.

[1] Curiously human blood has about the same salinity as sea water—a clue about our origins!

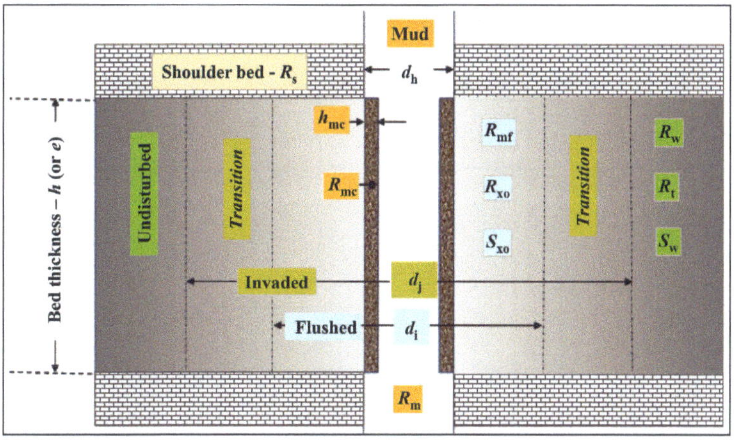

Figure 3.3 The borehole and surrounding zones.

Table 3.1 The logging environment[a].

Name of Zone	Dimension	Fluid Content	Fluid Resistivity	Rock Resistivity	Water Saturation
Bed	h (or "e")	-	-	R_t	-
Adjacent Bed	-	-	-	R_s	-
Borehole	d_h	Mud	R_m	-	-
Mudcake	h_{mc}	-	R_{mc}	-	-
Flushed	d_i	Residual Hydrocarbon & Mud Filtrate	R_{mf}	R_{xo}	S_{xo}
Invaded	d_j	Mixture	-	R_i	Transition xo → w
Uninvaded	-	Hydrocarbon & Connate Water	R_w	R_t	S_w

[a]In the case of a step profile invasion d_i and d_j are identical. Note also that d_i may also be referred to in the literature as D_i.

3.5 Electrode arrangements and depth of investigation

From the early days of surface prospecting the current and voltage electrode symbols (A B M & N) transferred from the horizontal to the vertical with the result that the basic electric logs were called either "normal" or "lateral." The distinction was a simple matter of where the current and measure electrodes were placed, either at surface or downhole mounted

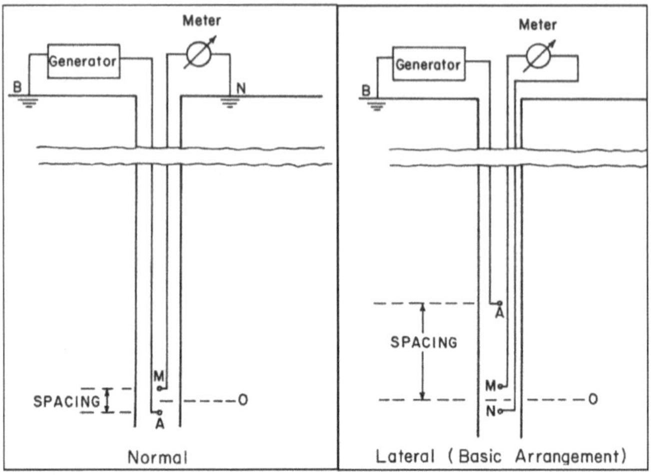

Figure 3.4 Electrode arrangements for "normal" and "lateral" devices.

on what amounted to a broomstick. The "normal" device is illustrated in Fig. 3.4 where electrical current is sent between a ground electrode B, at surface, and the A electrode, downhole. The resulting potential difference generated is then monitored between the M (downhole) and N (at surface) electrodes. The effective measure point is the midpoint between the A and M electrodes and the "spacing" is just the distance between A and M. Since there was no focusing, the current that did not leak away through the mud column flowed in a quasi-spherical zone of the surrounding formation with a diameter close to the AM spacing. Thus a 16″ SN device made most of its measurement of the formation resistivity within an invasion diameter of 16″. For that reason it was considered as a fair measure of R_{xo}.

Normal curves are called Short Normals when their AM spacing is between 5″ and 20″. Likewise long normals have the AM spacing between 20″ and 84″. The lateral devices have spacings of several feet. These different spacings were selected so that a fixed set of electrodes could be "re-wired" in turn to create the normal and lateral devices required without altering the downhole hardware. Fig. 3.5 illustrates this concept.

By use of the four electrodes (I, II, III, and IV), the early loggers could perform all three formation resistivity measurements using a four-conductor cable. The *lateral* logging system placed the A, M, and N electrodes downhole with the measure point between M and N and the *spacing* from A to the midpoint between N and M. It is instructive to see the results of some early experimental spacings which are shown in Fig. 3.6. A resistive bed, about 30′ thick, was successively logged by four

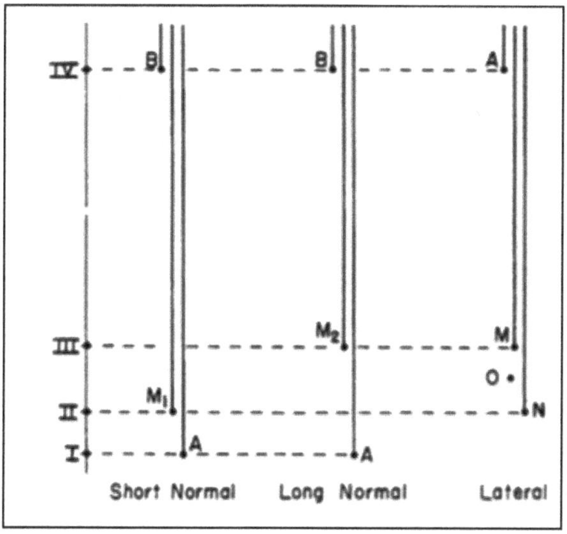

Figure 3.5 Electrode arrangement for normals and laterals. *Image courtesy Schlumberger (From Document 4).*

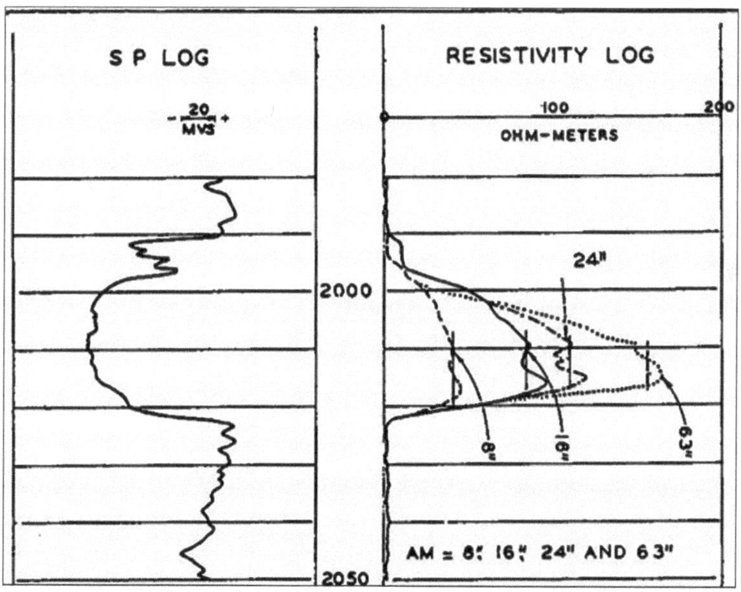

Figure 3.6 Effect of AM spacing on radial depth of investigation.

different normal devices with AM spacings of 8″, 16″, 24″, and 63″[2]. It is noticeable that the formation is invaded with a salty mud filtrate causing the near-wellbore R_{xo} to be less than R_t. The longer the AM spacing the deeper the *normal* device reads radially into the formation.

3.6 Very early example log

Fig. 3.7 shows a log made in 1936 in the Leesville field for The Texas Co. It is instructive to note that there are only two tracks. The left-hand track is labeled "porosity log" and is scaled in millivolts—evidently a spontaneous potential (SP) log. The other track is labeled "resistivity log" and is scaled in $\Omega m^2/m$ but makes no mention of the electrode arrangement or the spacing.

3.7 The resistivity exclusion principle

In the business world it is popular to link the cost, quality, and availability of a product using a three-by-three matrix. Ideally one would hope to create a product that was of good quality, that could be designed and built quickly and at low cost. Unfortunately, more often than not, one is able to control only two of the three variables. The matrix below shows that if the product is to be of excellent quality and made available quickly it will be expensive. If it is quickly available and cheap then it will be of poor quality. Finally, if the product is to be of the finest quality and cheap then we may have to wait for a long, long time until it becomes available.

Speed	**Quality**	**Cost**
Slow	Good	Expensive
Fast	Poor	Cheap
Slow	Good	Cheap

A similar situation arises when attempts are made to analyze old logs for the elusive number to use for R_t, the undisturbed formation resistivity. Today we are spoiled with sophisticated resistivity tools that define formation resistivity, both near the wellbore and far from it, in all kinds of mud systems and in all bed thicknesses, from the super thin to blanket sands. In

[2] The choice of 8, 16, 20 and 63 inches is not arbitrary—these measures reflect the underlying metric measures of 20, 40, 60, and 160 cm!

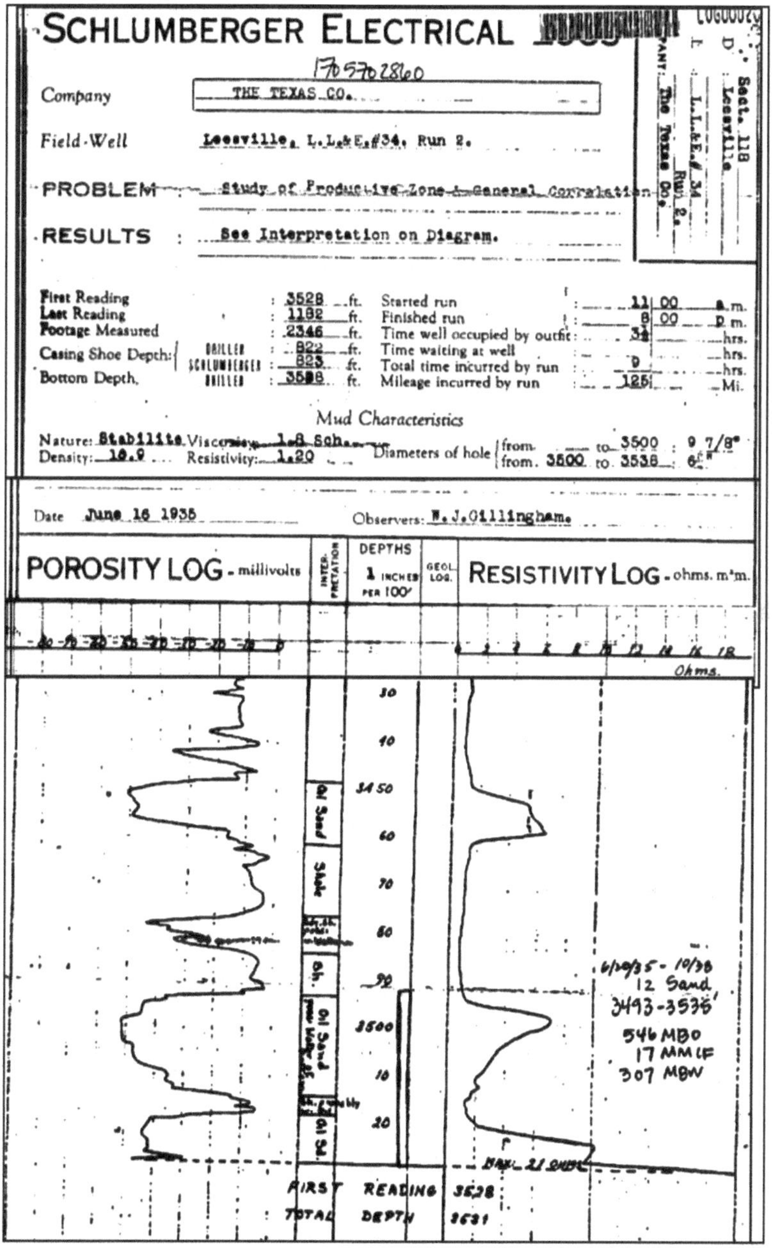

Figure 3.7 Very early electric log.

the pre-digital world we have the three interrelated variables as R_t, which is what we seek (to plug into Archie), bed thickness (h), and what might be called "ease of discovery." The three-by-three matrix for this system shows us that, at best, we can hope to discover a reasonable value for R_t with a reasonable effort only when the bed thickness is large.

Bed Thickness	Work Required	R_t
Thick	Reasonable	Acceptable
Thin	Very Difficult	Highly Questionable
Unknown	Monumental	Probably Unknowable

In thin beds the process for finding R_t is convoluted and involves the interplay of electrode spacings (normal or lateral) and their arrangements with respect to surrounding beds. Note that the terminology used on these pages uses the symbols:

R_m for mud resistivity (see log heading)

R_s for surrounding bed resistivity (also known as "shoulder bed" resistivity)

e for bed thickness (in ft, more commonly given the symbol *h* today)

$R_{16''}$ for SN resistivity

$R_{64''}$ for long normal resistivity

$R_{18'8''}$ for lateral resistivity

R_{LL3} for laterolog 3

R_{LL7} for laterolog 7

R_{ILd} for deep induction

R_{ILm} for medium induction

Conventional wisdom divides the process for finding R_t into three ranges of formation resistivity: Low, medium, and high. A useful guide is offered in Schlumberger's 1966 Chart Book, on pages B-7 and B-8, which offer a highly condensed form of the copious information contained in the Document 4, already mentioned. The procedures to be following may be summarized in the following section.

3.8 Finding R_t in low-resistivity zones

Low resistivity here is defined as cases where $R_{16''}/R_m$ is $<$ 10, then if bed thickness is 20′ or less R_t is close to $R_{64''}$ multiplied by the factor 20/*e*.

$$R_t = R_{64''} x \left(\frac{20}{e} \right)$$

This is shown in more detail on Fig. 3.8

BED THICKNESS (e)	QUALIFICATIONS		DEVICE	RESPONSE
A. IN LOW RESISTIVITY, WHEN $R_{16''}/R_m < 10$ (INVASION UP TO 2d)				
$e > 20'$ (>4 AM')			Long Normal	$R_{64''} = R_t$
$e \approx 15'$ (3AM')	$R_m \simeq R_s$	$R_{64''}/R_s \geq 2.5$	Long Normal	$R_{64''} = 2/3\ R_t$
$e \simeq 15'$ (3AM')	$R_m \simeq R_s$	$R_{64''}/R_s \leq 1.5$	Long Normal	$R_{64''} = R_t$
$e \simeq 10'$ (2AM')	$R_m \simeq R_s$	$R_{64''}/R_s \geq 2.5$	Long Normal	$R_{64''} = 1/2\ R_t$
$e \simeq 10'$ (2AM')	$R_m \simeq R_s$	$R_{64''}/R_s = 1.5$	Long Normal	$R_{64''} = 2/3\ R_t$
$5' < e < 10'$	When oil bearing and SP is -50 – 80 MV		Short Normal	$R_{16''} \simeq R_t$
$5' < e < 10'$	Surrounding beds homogenous		Lateral in resistive bed	$R_t \lesssim R_{Max} \times R_s/R_{Min}$
Thin beds (in general)	Surrounding beds homogenous		Lateral in conductive bed	$R_{19''} \simeq R_t$

Figure 3.8 Corrections for long and short normals to R_t in "low"-resistivity zones. Courtesy Schlumberger.

3.9 Finding R_t in medium-resistivity zones

Medium resistivity is defined as cases where $10 < R_{16''}/R_m < 50$. In the general case use $R_t = R_{64''}$.

However, if the bed thickness is 20′ or greater, then R_t should be taken from the $R_{18'8''}$ Lateral multiplied by a factor between 1 and 0.67.

$$R_t = R_{18'8''} x \text{ (factor picked from chart—Fig 3.9)}$$

If bed thickness is 10′ or less, then:

$$R_t = R_{Max} x \left(\frac{R_s}{R_{Min}} \right)$$

As shown in more detail in Fig. 3.9.

3.10 Finding R_t in high-resistivity zones

High resistivity is defined as cases where $R_{16''}/R_m > 10$. Where $R_{16''}/R_{64''}$ is less than 1, R_t is calculated as $(R_{64''})^2/R_{16''}$ but if the $R_{16''}/R_{64''}$ is greater than 1 then $R_{18'8''}$ is the best choice, provided the requisite adjustments are made for thin beds following Fig. 3.9.

A convenient way to summarize these somewhat complicated guidelines for R_t from old E-logs is given in the following Fig. 3.10. A good practical staring point is to read the basic resistivity numbers including R_m, $R_{16''}$, $R_{64''}$, $R_{18'8''}$, and R_s and enter them on the top left-hand corner panel of the Table 3.2. The next step will be to compute the ratios required in the right-hand panel, namely $R_{16''}/R_m$, $R_{64''}/R_s$, and

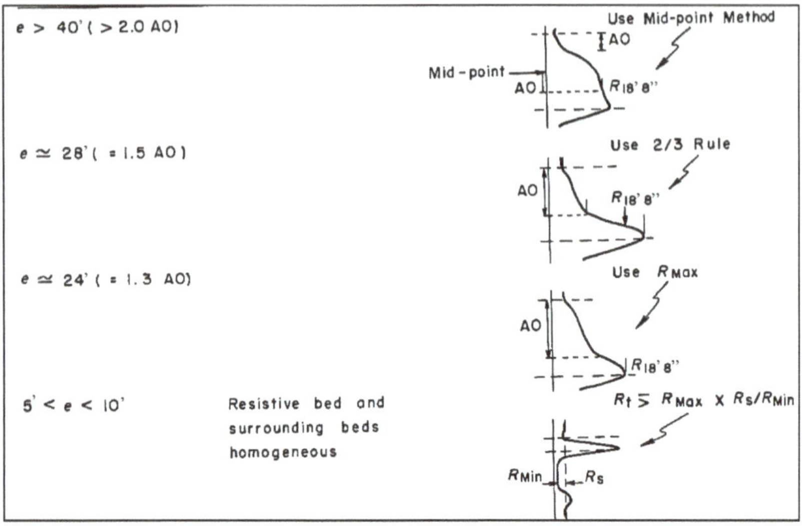

Figure 3.9 Corrections for 18′8″ lateral to R_t.

Resistivity readings Ωm			Ratios		Color code	
Mud	R_m		$R_{16''}/R_m$		10 –	10 +
Short normal	$R_{16''}$		$R_{64''}/R_s$		2.5 –	2.5 +
Long normal	$R_{64''}$		$R_{16''}/R_{64''}$		1 +	1 –
Lateral	$R_{18'8''}$					
Shoulder bed	R_s					

$R_{16''}/R_m$ Code				
Low res formation			High res formation	
Bed thickness Ft	$R_{64''}/R_s$ code	R_t	$R_{16''}/R_{64''}$ code	R_t
20+		$R_{64''}$		$(R_{64''})^2/R_{16''}$
				$R_{18'8''}$
15–20		$R_{64''}$		
		$R_{64''} \times 1.5$		
10–15		$R_{64''} \times 1.5$		
		$R_{64''} \times 2$		
10 –		$R_{16''}$		

Figure 3.10 Guide to R_t.

Table 3.2 Invasion response of laterologs-3 and -7.

di"	R_{LL7}	R_{LL3}
20	$0.2\ R_{xo} + 0.8\ R_t$	$0.25\ R_{xo} + 0.75R_t$
40	$0.4\ R_{xo} + 0.6\ R_t$	$0.5\ R_{xo} + 0.5R_t$
80	$0.6\ R_{xo} + 0.4\ R_t$	$0.75\ R_{xo} + 0.25\ R_t$

$R_{16''}/R_{64''}$. Depending on the $R_{16''}/R_m$ ratio being greater or less than 10 the color code orange or yellow leads to one of the two columns in the lower panel of the guide labeled "Low Res Formation" or "High Res Formation." On the left-hand (Low Res) side the bed thickness column is divided into four segments ranging from less than 10′ to greater than 20′. For each of these there are two choices given, a green or a blue. The green coded choice corresponds to an $R_{64''}/R_s$ ratio of less than 2.5 and the blue to more than 2.5. The value to use for R_t is then given as a function of $R_{64''}$ or $R_{16''}$. For the High Res Formation (right-hand) side of the lower panel the $R_{16''}/R_{64''}$ ratio determines if R_t is derived from the 18′8″ lateral or as a function of $R_{64''}$ and $R_{16''}$. Note that Table 3.2 leaves blank spaces for the analyst to note the actual log readings.

3.11 Use of laterologs for R_t

The laterologs-3 and -7 were introduced as focused devices that could work far better than conventional electric logs in salty mud environments. As such their measure currents passed through any conductive mud column directly into the invaded zone before registering any response to the uninvaded zone. Fig. 3.10 offers some guidance to correcting the laterologs for invasion effects. Such corrections depend on the knowledge of the diameter of invasion, d_i, which, unfortunately, remained an unknown until the introduction, much later, of the dual laterolog tools that incorporated a μSFL measurement of R_{xo} that then allowed an estimate of d_i and the correction of a dual laterolog measurement to R_t. So, for the analysts faced with an LL3 or LL7, the only comfort that can be offered is a simple rule of thumb.

$$R_t = \frac{R_{LL} - \alpha R_{xo}}{(1 - \alpha)}$$

With $\alpha = \frac{\sqrt[2]{d_i}}{15}$ for the LL7 and $\alpha = \frac{\sqrt[2]{d_i}}{12}$ for the LL3
Alternatively Table 3.2 can be used.

3.12 Origins of the resistivity correction charts

Many of the examples shown here were based on simulations made on an analog "resistor network" pseudo computer that was designed to imitate multiple combinations of all the variables of d_h, R_m, e, R_{xo}, R_t, d_i and what a 16″ or 64″ normal or a 18′8″ lateral device would read given the multiple inputs. Some of the expected aberrations can be illustrated here.

Bed boundaries may be missed or wrong and resistivities appear too low depending on the relative thickness of the bed of interest with respect to the electrode AM spacing. Fig. 3.11(a) gives an example of a bed which is six times thicker that the borehole diameter ($e = 6d$). The shoulder beds have the same resistivity as the mud ($R_s = R_m$). R_t is 14 Ωm but the maximum registered at the middle of the bed by the normal device is just over 9 Ωm. The bed thickness, as indicted by the inflection points, *P*, indicate a bed thinner than it actually is.

A thinner version of the same 14 Ωm formation, with $e = d$ reads slightly under 1 Ωm where it should read 14. In this case the AM spacing is greater than the bed thickness. The case of the resistivity device effectively becoming a "blunt instrument."

Lateral devices are not immune to these problems either. Depending on relative size of bed and electrode spacings bed boundaries are generally fine but resistivities are too low in thin beds and slightly high in thick ones. Of course the lateral device also requires water-based mud systems and should not be confused with laterolog devices that require salt mud systems for optimum performance. Other things being equal one could say, to a good approximation, that a 16″ SN is close to R_{xo} and an 18′8″ lateral is closest to R_t. However, neither is perfect. In particular the lateral devices suffer from "blind zones" and "reflection peaks." Fig. 3.12 (a) illustrates the distortions to be expected from a lateral device that read 20 Ωm in a 14 Ωm bed and exhibits a nonsymmetrical trace as it logs through a bed of uniform resistivity.

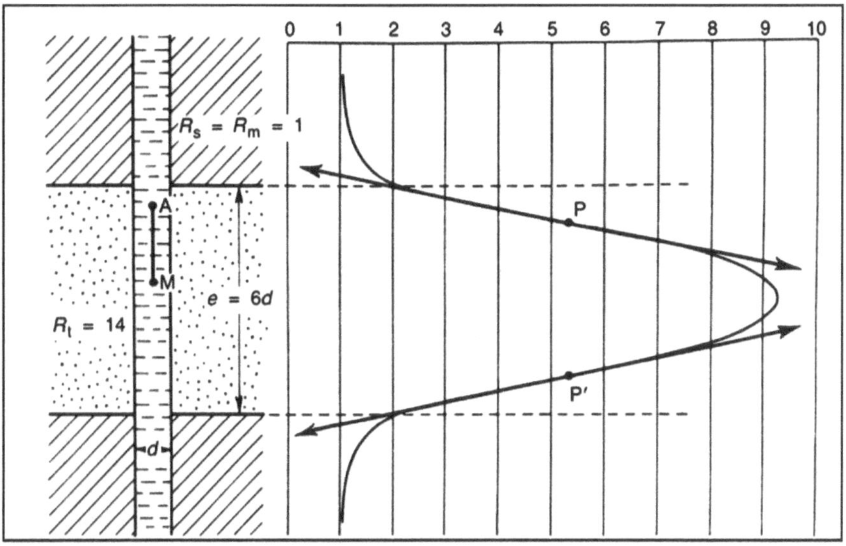

(a)

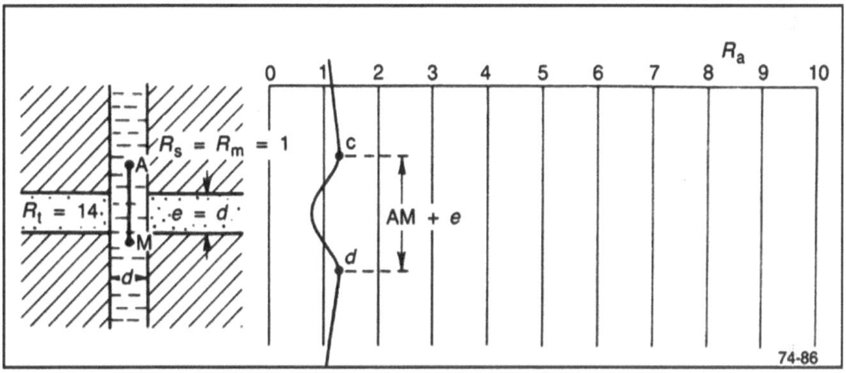

(b)

Figure 3.11 (a)—Normal device reads both R_t and bed thickness too low. (b)—Normal device "blind" to R_t and bed thickness too large ($c \rightarrow d$). Image courtesy Schlumberger.

Fig. 3.12 (b) shows the equally distortive trace produced when the lateral device logs though a thin bed where the same 14 Ωm R_t wobbles between 0.4 and 3.6 Ωm.

No clear bed boundaries are evident in the bed that has $e = 3$d. The behavior is characterized by a "blind zone" and a "reflection peak."

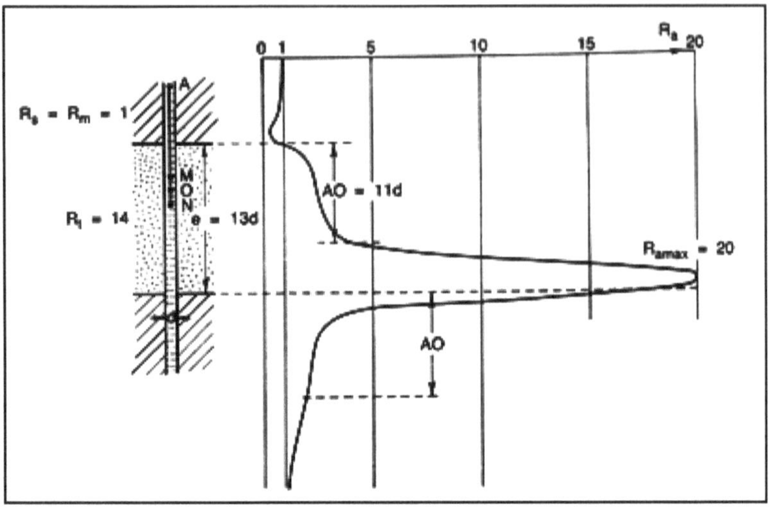

(a)

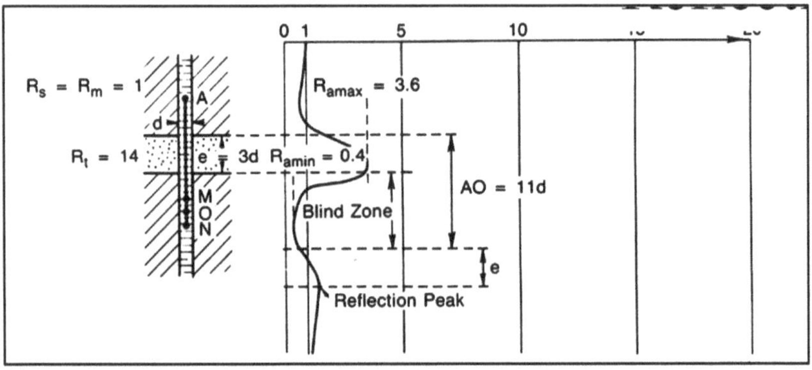

(b)

Figure 3.12 (a)—Lateral device distortions in thick bed. (b)—Lateral device distortions in thin bed. Image courtesy Schlumberger.

3.13 Porosity from the limestone lateral

The limestone lateral was used in low porosity, very high resistivity, carbonate formations. Its introduction was spurred on by the failure of conventional electric logs of the time to properly identify thin bed boundaries in cases where there was a large contrast between the mud resistivity and the formation resistivity. Such was the case in, for example in Western Kansas, where Paleozoic limestones lie under hundreds of feet of rock salt. The only way to successfully drill these formations was to allow the mud to reach the salt saturation. This resulted in salt saturated mud and a contrast between R_t and R_m of 1000:1. Similar conditions prevailed in the Permian Basin in West Texas and New Mexico. The solution was to avoid current leakage into the mud column. One solution, to come later, was the Laterolog with focusing (or "guard") electrodes. But first the solution used was the symmetrical lateral, known thereafter as "the limestone lateral." These devices had a symmetrical electrode array, as shown in Fig. 3.13, with AO spacings of 32″ or 37½″.

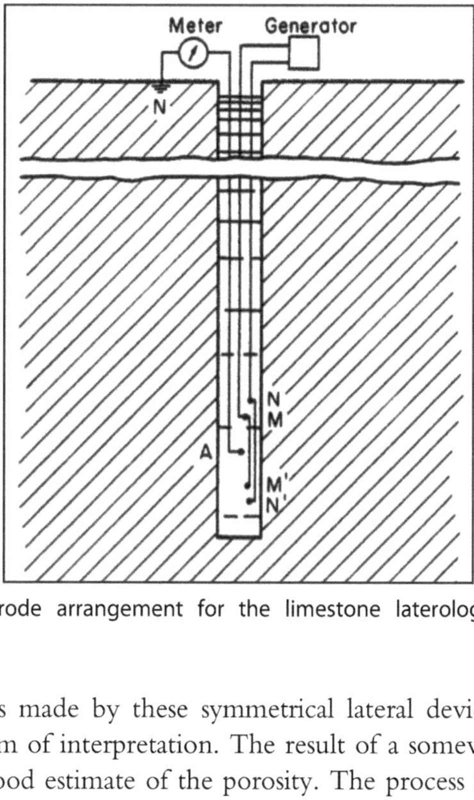

Figure 3.13 Electrode arrangement for the limestone laterolog. Image courtesy Schlumberger.

The readings made by these symmetrical lateral devices demanded a specialized system of interpretation. The result of a somewhat convoluted process was a good estimate of the porosity. The process is divided into a number of steps.

Step 1 calls for a value for the "maximum deflection" which is read from the Table 3.3. This requires inputs of both hole size and the AO spacing of the tool in use. By way of an example calculation we may use a data set with $R_a = 50\ \Omega m$, $R_m = 0.5\ \Omega m$, SP = −25 mV, hole size = 7 7/8″ and AO = 32″. Working with the given data, the R_a/R_m ratio for a 7 7/8″ hole and an AO of 32″ is 128 Ωm. The "maximum deflection" is thus given by $R_m \times R_a/R_m = 0.5 \times 128 = 64\ \Omega m$.

Table 3.3 Lookup for max deflection.

	Max. Ra/Rm	
HOLE SIZE	AO = 32"	AO = 37 1/2"
4 3/4"	445	605
6 3/4"	177	232
7 7/8"	128	171
8 3/4"	102	141

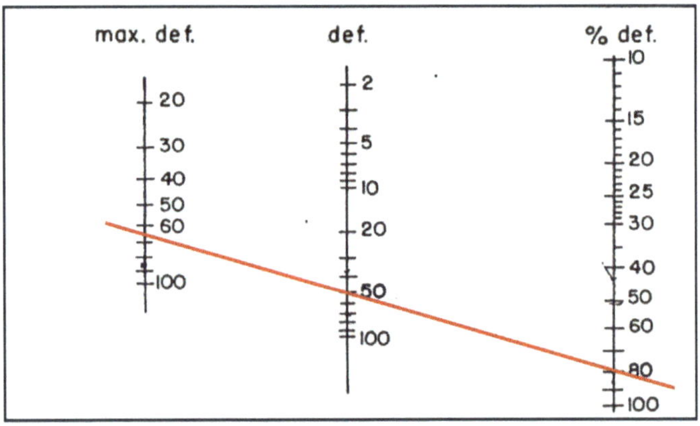

Figure 3.14 Nomogram to determine % deflection.

Step 2 seeks the "% deflection" which is found from the nomogram of Fig. 3.14. It is instructive that there was a need in the 1950s for such a simple nomogram which takes two numbers and divides one by the other. In this case 50/64 ≈ 80%. In those days there were no handheld calculators and the average log analysts probably did not carry a slide rule to the well site!

Step 3 takes the % Deflection number, 80% in this case, and enters it on the chart shown in Fig. 3.15. At the intersection of the red line (80%) and the green line (7 7/8″ hole size) the porosity index (PI) is read as 5.5 by interpolation between the lines for 4 and 6.

The final step converts the PI into porosity by multiplying the PI by a coefficient "*C*" that, in turn, depends on whether the formation is oil or water bearing and also on the SP deflection. "*C*" is determined from Table 3.4. In this case, with an SP of −25 mV, $C = 1.3$ which is found

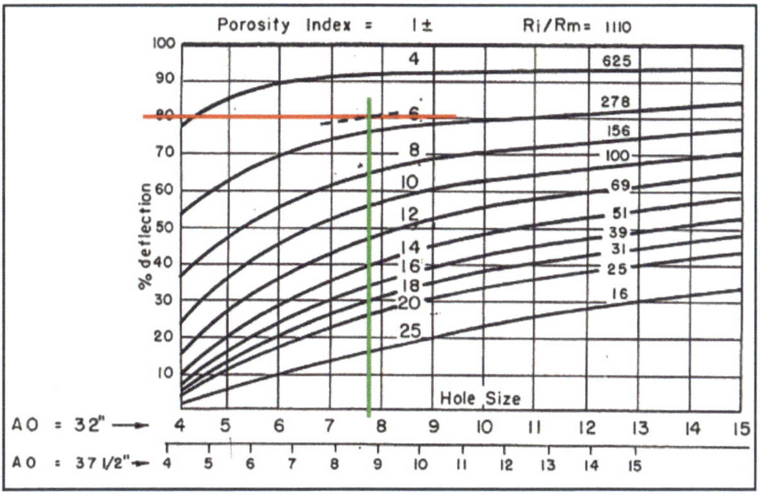

Figure 3.15 Chart for porosity index.

Table 3.4 Coefficient "C" for porosity from PI.

WATER – BEARING FORMATIONS		OIL – BEARING FORMATIONS	
WHEN	THEN	WHEN	THEN
" SP = 0	" C = 1	" SP = 0	" C = 1.4
" SP = –50 MV	" C = 0.86	" SP = – 50 MV	" C = 1.22
" SP = – 100MV	" C = 0.7	" SP = – 100 MV	" C = 1

by interpolation between the 1.22 and 1.4 values. Thus, at the end of this convoluted chain, we find that the porosity is $1.3 \times 5.5 = 7.2\%$.

A sample recording with the limestone lateral is shown in Fig. 3.16.

3.14 Induction logs

Induction logs are familiar to today's analysts, so we will only touch on them in so far as they were used in the pre-digital period for purposes and in ways that differ from modern practice. Their characteristics include being more focused than regular E-logs, that they work in air filled holes and oil-base muds, that they have poor bed resolution (±6 ft for the

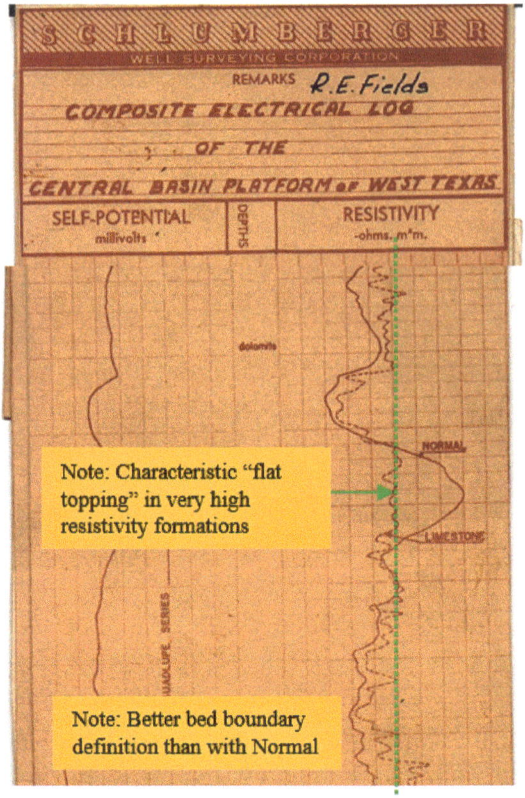

Figure 3.16 Limestone lateral example (San Andres Formation).

early 5 and 6FF40 tools) and readings obtained in salty mud systems required objectionably large environmental corrections. Fig. 3.17 gives the layout and format of a typical early induction log.

Things to notice are that the basic measurement, which is formation conductivity, is shown on the log in the third track. This is a linear conductivity scale in millimhos. The resistivity in the second track is a "reciprocated" version of the conductivity measurement.

Accompanying the induction resistivity trace in track two is a linear SN with amplified backup traces. A millimho is 1/1000th of a mho, which in turn is the reciprocal of an ohm. This 0 conductivity corresponds to an infinite resistivity, 10 mmhos to 100 Ωm, 100 mmhos to 10 Ωm, and 1000 mmhos to 1 Ωm.

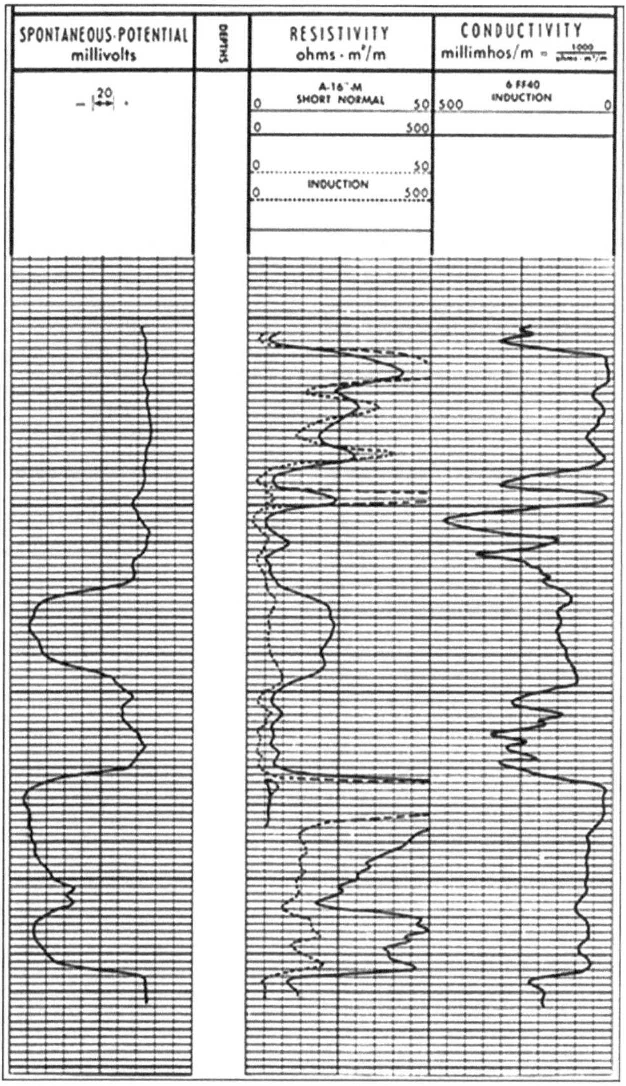

Figure 3.17 Induction log presentation. Image courtesy Schlumberger.

One of the first "combination tools" was the Induction-Sonic which allowed a simultaneous recording of both resistivity and porosity data with just one logging run in the hole. This opened up the field to making wellsite "log analysis" in the basic form of generating an R_{wa} curve based

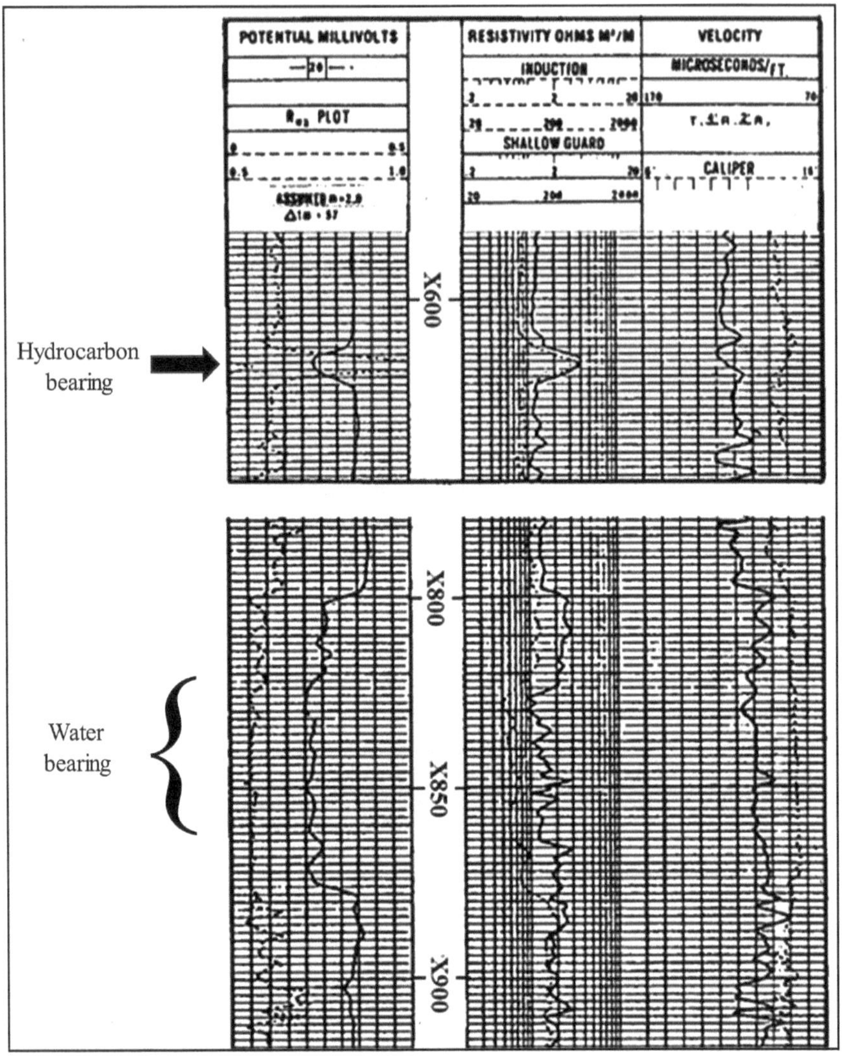

Figure 3.18 Combination induction-sonic log with R_{wa} curve. Image courtesy Schlumberger.

on deep formation resistivity from the induction tool (R_t) and porosity from the sonic tool via the transform of Δt to ϕ_s. Fig. 3.18 gives an example of this presentation[3].

[3] Chapter 13 goes into more detail on these and other similar "Quick-look" approaches to wellsite analysis in the pre-digital age.

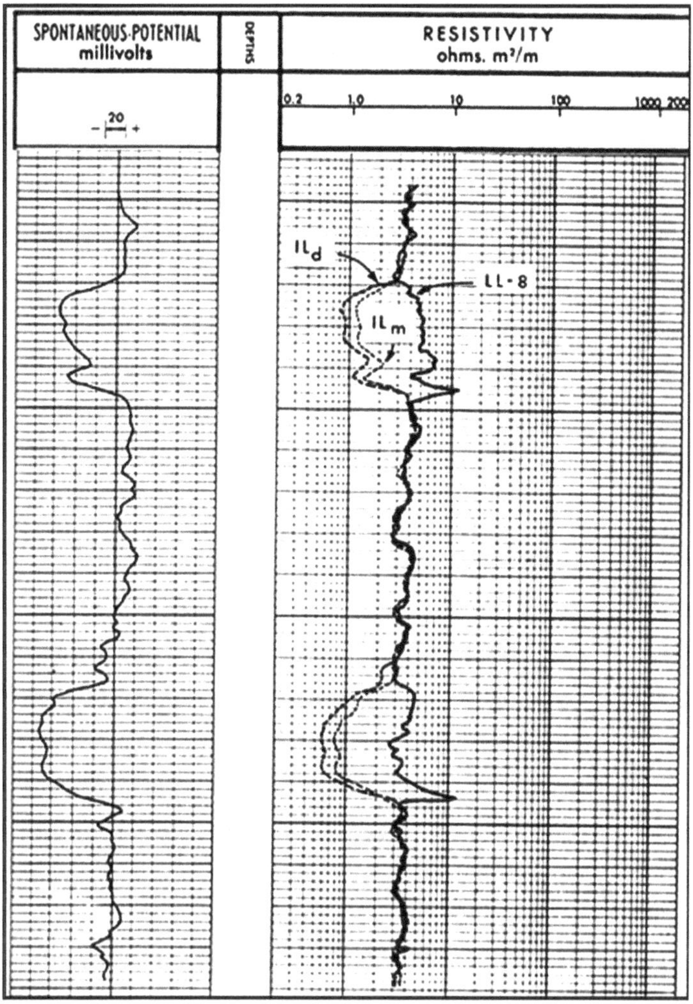

Figure 3.19 Dual induction LL8. Image courtesy Schlumberger.

3.15 Dual induction logs

Dual Induction tools, introduced in 1963, combined a Deep Induction (ILd) with a Medium Induction (ILm) and a variety of near-wellbore shallow resistivity devices such as the SN, the Laterolog-8 (LL8), or the SFL. The purpose of the simultaneous triple resistivity measurement was to allow a solution to the challenge of defining R_t, R_{xo}, and d_i. With three unknowns, three measurements are required to find a solution. Today's analysts will be familiar with the "tornado chart" that gives a graphical solution to the determination of R_t, R_{xo}, and d_i. Fig. 3.19 gives an example of a Dual Induction LL8.

Modern multispacing induction tools can go even further and "map" the invasion profile pin-pointing such things as invasion annuli if, and when, they develop. However, some of the older invasion charts may not be familiar to the reader. Fig. 3.20 shows an early "Tornado Chart" that could be used with the log shown in Fig. 3.19.

The raw log readings for the data plotted on Fig. 3.19 are: $R_{ILd} = 1.2$, $R_{ILm} = 1.9$, and $R_{LL8} = 9.5$. These give rise to the ratios of: $R_{LL8}/R_{ILd} = 7.9$ and $R_{ILm}/R_{ILd} = 1.58$. At the intersection of the respective lines from the X- and Y-axes the ratio of R_t/R_{ILd} is read as 0.83 and the ratio of R_{xo}/R_t as 13. From which we may conclude that $R_t = 0.83 \times R_{ILd} = 0.83 \times 1.2 = 1.0$ and $R_{xo} = 13 \times R_t = 13 \times 1 = 13$. d_i is read directly from the chart as 52″.

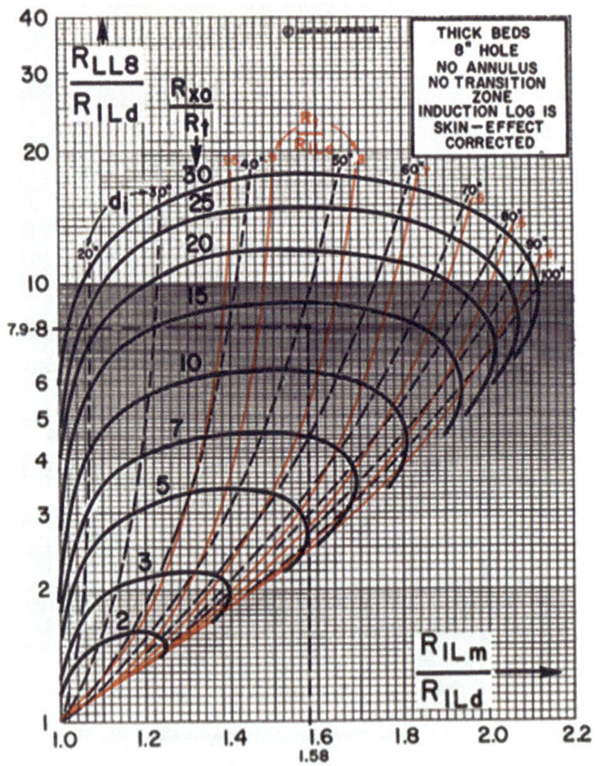

Figure 3.20 Early tornado chart. Image courtesy Schlumberger.

3.16 Dual Laterologs

For salty mud systems the Dual Laterolog was introduced in 1972 and, again, offered two laterolog measurements (deep and shallow) as well as a near-wellbore resistivity measurement. An example of a dual laterolog is shown in Fig. 3.21.

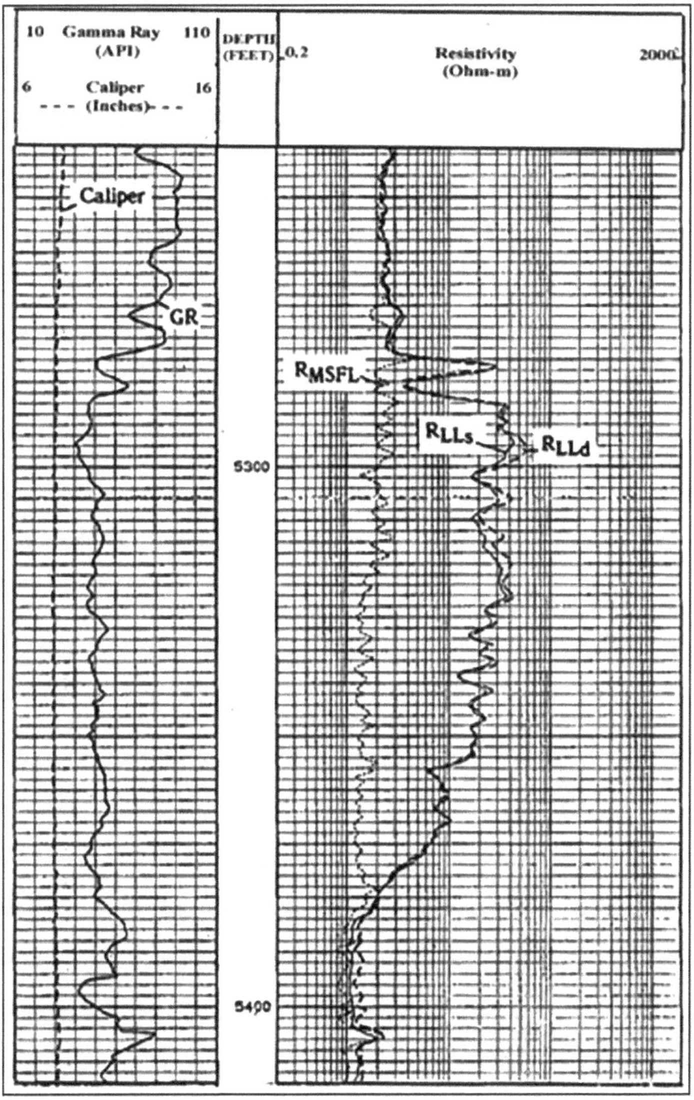

Figure 3.21 Example of Dual Laterolog-MSFL.

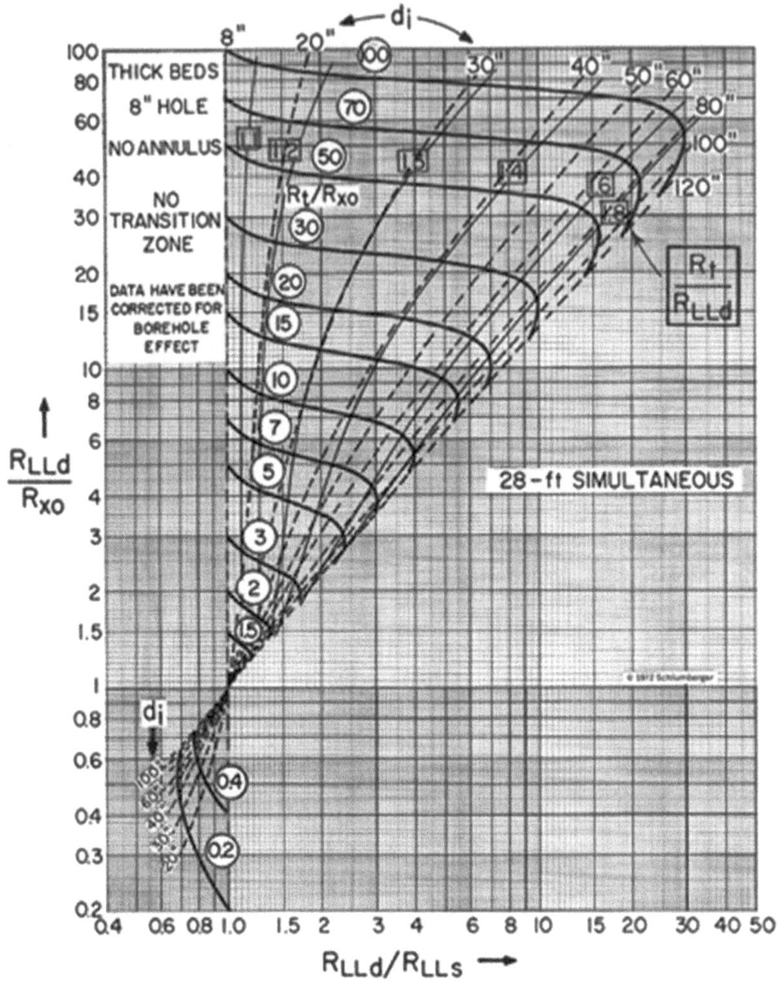

Figure 3.22 Dual Laterolog "Butterfly" chart. Image courtesy Schlumberger.

Figure 3.22 shows the Dual Laterolog "Butterfly" chart, used to determine R_t, R_{xo} and d_i from the three measurements made by the Dual Laterolog combination tool. It serves the same purpose as the "Tornado" chart does for the Dual Induction tool, already discussed.

3.17 Micro-resistivity devices

Initially the microlog served as a detector of presence of mudcake which in turn was taken to indicate that a porous and permeable formation had

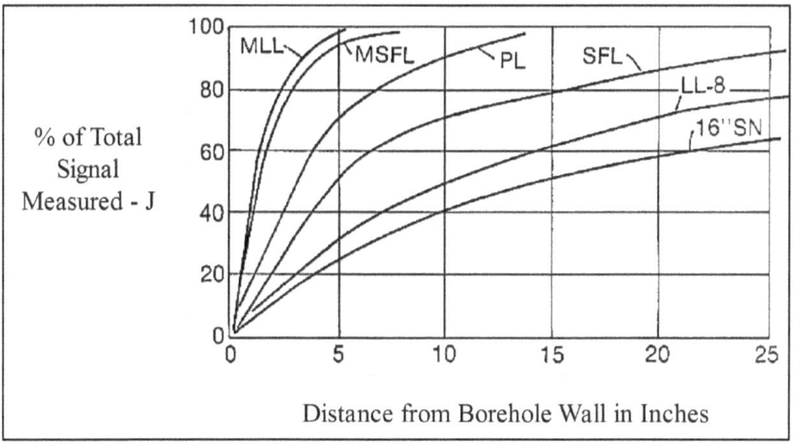

Figure 3.23 Depth of investigation of micro-resistivity devices.

been drilled through. Chapter 8 explains in detail the working and interpretation of the Microlog measurements for porosity and invaded zone, near well bore resistivities and will not be repeated here. What is of relevance that following the introduction of the Microlog a number of other devices became popular and were used as indicators of what the analysts hoped was R_{xo}. Each of these many devices had its characteristic "depth of investigation." Since the reader will undoubtedly come across some or all of these near well bore tools it is useful to have a guide to which one reads closest to R_{xo} and which is picking up a goodly proportion of R_t as well. Fig. 3.23 gives the "pseudo geometrical factors" of five of the most common ones: Microlaterolog, Proximity log, LL8, LL3, and LL7[4].

If the values of R_{xo} and R_t and the diameter of invasion, di, are known then the chart can be used to calculate the response of any one of the five tools shown by applying the following formula:

$$R_a = JR_{xo} + R_t(1 - J)$$

For example the response of the LL8 when $R_{xo} = 1$, $R_t = 10$ and $d_i = 40''$ is given by

$$R_{LL-8} = 0.8 \times 1 + 0.2 \times 10 = 2.8$$

A better choice for measuring R_{xo} in this case would be the proximity log which would read just under 1.4. Of course in the real world the

[4] The unfocused 16" short normal is not included on the chart since it is not directly comparable to the others which are focused tools.

Table 3.5 Comparative metrics for micro devices.

Tool Name	Depth for 90% Response Inches	Bed Resolution Inches
Microlog	1	3
Microlaterolog	4	6
Microsperically Focused Log	4.5	12
Proximity Log	10	12
Sperically Focused Log	24	12
Laterolog 8	49	12
16" Short Normal	70+	24

values of R_{xo}, R_t, and d_i are unknown and the search for R_t is more complicated but at least knowing the relative depths of investigation of the "micro" devices will help in making a better-informed guess at the true value of R_t.

Of help in this respect is the Table 3.5 which ranks the micro devices by "depth to 90% response" and by "bed resolution." The table also includes the 16″ SN that can now be seen more clearly to be a relatively deep investigation device.

Further reading

Archie, G.E., 1942. The electrical resistivity log as an aid in determining some reservoir characteristics. J. Pet. Technol. 146, 54–62.

Doll, H.G., 1949. Introduction to induction logging. J. Pet. Tech. 1 (6).

Doll, H.G., 1951. The laterolog; a new resistivity logging method with electrodes using an automatic focusing system. Petr. Trans. AIME. 192, 305–316.

Doll, H.G., 1953a. Two decades of electrical logging. J. Pet. Technol. 5 (9), 33–41.

Doll, H.G., 1953b. The microlaterolog. J. Pet. Tech 5 (1), 17–32.

Doll, H.G., Martin, M., DDDD: Electrical logging in limestone Fields. Proceeding of the Third World Petroleum Congress, 28 May–6 June 1951, the Hague, the Netherlands

Guyod, H., 1944. Electrical well logging: fundamentals. Oil Weekly 114 (7/8/44) and 115 (25/12/45). nol., 1 (6), 211–219.

Guyod, H., 1964. Factors affecting the responses of laterolog-type logging system (LL3 and LL7). J. Pet. Tech 16 (2).

Hamilton, R.G., 1960. Application of the proximity log. Trans. SPWLA, 1st Ann. Log. Symp.

Martin, M., Murray, G.H., Gillingham, W.J., 1938. Determination of the potential productivity of oil-bearing formations by resistivity measurements. Geophysics 3 (3), 252–272.
Owen, J.E., Greer, W.J., 1951. The guard electrode logging system. Petr. Tech. 3 (12).
Schlumberger Ltd, 1970. The Dual Laterolog, Technical Report. Schlumberger Ltd.
Schlumberger, C., Schlumberger, M., 1929. Electrical logs and correlations in drill holes. Pet. Eng. 86, 87, 90.
Schlumberger, C., Schlumberger, M., Leonardon, E.G., 1934. Electrical coring: a method of determining bottom-hole data by electrical measurements. Trans. AIME 110, 237 (1934).
Schlumberger Well Surveying Corp, 1947. Resistivity Departure Curves. Schlumberger Well Surveying Corp.
Schlumberger Well Surveying Co, 1947. Resistivity Departure Curves, Document 3. Schlumberger Well Surveying Co.
Schlumberger Well Surveying Co, 1950. Interpretation Handbook for Resistivity Logs. Document 4. Schlumberger Well Surveying Co.
Schlumberger Well Surveying Co, 1951. Interpretation of Induction Logs. Schlumberger Well Surveying Co.
Schlumberger Well Surveying Co, 1962. Induction Log Correction Charts. Schlumberger Well Surveying Co.
Tixier, M.P., 1951. Porosity index in limestone from electrical logs. Oil Gas J.
Tixier, M.P., Alger, R.P., Biggs, W.P., Carpenter, B.N., 1963. Dual induction-Laterolog: A new tool for resistivity analysis. Soc. Petrol. Eng. of AIME Pap. 713, 6-9 (October).
Tixier, M.P., Alger, R.P., Biggs, W.P., Carpenter, B.N., 1965. Combined logs pinpoint reservoir resistivity. Petrol. Eng. 13, 212–229.

Appendix

Introduction to Schlumberger Document 4

INTRODUCTION

Resistivity measurements, together with measurements of spontaneous potentials, constitute the essential elements of the electrical logging of bore holes. After more than fifteen years of universal practice in the oil industry, it is now commonly known that these measurements provide a remarkable record of all the formations traversed, and give valuable information regarding the fluid content of the reservoir rocks.

It has not therefore seemed necessary to include in the present book demonstrations of the field of application and of the efficiency of resistivity logs. The book is rather intended to be a guide for the interpretation of these logs.

A resistivity log usually comprises several curves recorded in terms of depth. These curves show numerous peaks, depressions, and flat zones corresponding to the various resistivities of the formations traversed by the bore hole.

As the electrode system passes along a bore-hole through material of one resistivity into a material of different resistivity, this passage appears on the graph not as an abrupt step exactly opposite the interface between the materials, but as a transition which extends a short distance above and below the interface. Moreover, the maximum and minimum values shown on the logs opposite each formation are often different from the true values of the resistivities. This difference exists because of the influence on the measurements of factors other than the resistivities of the formations; such as, physical characteristics of the electrode system, mud resistivity, hole diameter, bed thickness, ...

The influence of these different factors upon the shape and the magnitude of the deflections of the resistivity curves will be analyzed in the present paper.

The discussion is based on mathematical computations and on laboratory tests, confirmed by an extensive field experience. Theoretical developments have been, however, eliminated from the text of the book and confined, in summarized form, to an appendix. The paper is limited essentially to the explanation of the results, and to their confrontation with practical problems. Efforts were made to bring forward some rules of interpretation and to show the application of these rules to actual examples.

It is hoped that the conclusions of this review will help the users of electrical logging to derive, from the method, more accurate and more complete information.

Chapter I contains general considerations on electrical resistivity of earth formations, and on the essential characteristics of the measuring devices used for the recording of resistivity in bore holes. Some consideration is also given to the anisotropy of sedimentary strata.

Chapter II is devoted to the characteristic shapes of resistivity curves recorded with the conventional measuring devices, in relation with the determination of the boundaries of geological formations by means of resistivity logs.

Chapter III, Chapter IV and Chapter V deal with the question of the investigation of true resistivity and discuss the results derived from the departure curves.

Chapter VI is related to the investigation of fluid content in reservoir rocks.

Chapter VII is a review of some actual examples of interpretation.

Images courtesy Schlumberger

Table of Contents of Schlumberger Document 4

CONTENTS

CHAPTER III

INVESTIGATION OF TRUE RESISTIVITIES OF FORMATIONS — CASE OF BEDS OF INFINITE THICKNESS

CHAPTER IV

INVESTIGATION OF TRUE RESISTIVITY (CONTINUED) — CASE OF BEDS OF FINITE THICKNESS

CHAPTER V

PRACTICAL CONCLUSIONS ON TRUE RESISTIVITY INVESTIGATION

SECTION III

Sources for porosity

CHAPTER 4

Sonic logging for porosity and lithology

Abstract

The sonic logging tools hold a special place in the arsenal of the formation evaluator. At the same time, measurement of the propagation of a compressional wave through a porous medium reflects on a plethora of information about the subsurface, probably greater than from any other single device. The analyst can use it to define porosity, lithology, compaction, overpressure, and fractures and even indicate organic content of shales. Hopefully, this chapter will help today's analyst to fruitfully use the old sonic logs, sleeping peacefully, in those file cabinets—but only waiting to be re-awakened for useful purpose again.

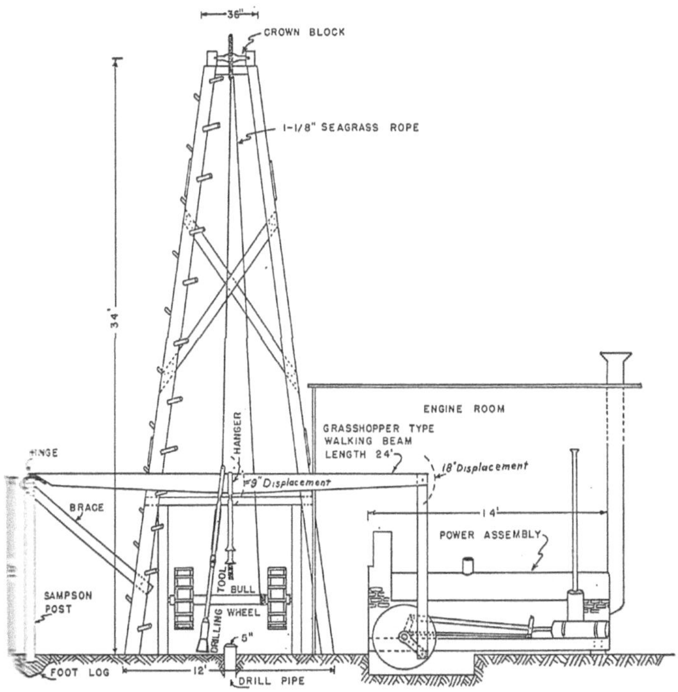

THE DRAKE RIG

Courtesy of "The Wildcatters" by Samuel W. Tait, Jr.

Formation Evaluation with Pre-Digital Well Logs
DOI: https://doi.org/10.1016/B978-0-12-820232-6.00004-2

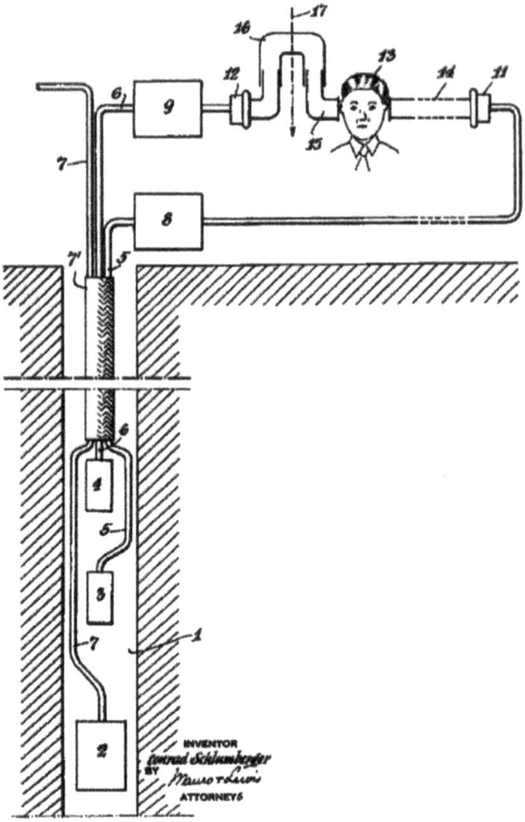

Figure 4.1 First attempts at sonic logging. *Reproduced from French Patent 786.863, 1935.*

4.1 Fundamentals

The earliest attempt to measure the speed of sound waves in a borehole was made by Conrad Schlumberger in 1935, and he set Henri Doll to conduct experiments using a horn from a Model A Ford and two geophones. The setup is shown[1] in Fig. 4.1. The story goes that Doll's assistants wired in a microphone to Doll's headphones, and when he let off loud blasts from the horn, they whispered "Trilobites speaking. Go to hell[2]."

[1] From the corresponding patent application.
[2] Quoted from "Schlumberger—The History of a Technique"—p. 309.

Table 4.1 Speed of sound.

	MPH	Ft/sec	μsec/ft
Air	768	1,126	888
Water	3,608	5,291	189
Sandstone	12,276	18,000	56

4.2 The speed of sound

The speed of sound offers a convenient gauge of the "stiffness" of the materials it passes through, going fastest in solids (such as reservoir rocks), less quickly in liquids (such as oil or water), and slowest of all in gases (such as methane or air).

Table 4.1 gives a useful summary. Early work with core samples suggested that a relatively simple relationship might exist between the speed of sound through a porous medium and its porosity. MRJ Wyllie published his time average equation in 1956. He worked in units of the *reciprocal* of the speed of sound or, in the jargon of the log analyst, the *slowness*, which was given the symbol Δt and measured in microseconds per foot (or meter) with the symbol μs/ft. His equation states:

$$\phi = \frac{\Delta t - \Delta t_{ma}}{\Delta t_{fl} - \Delta t_{ma}}$$

This simple equation may also be viewed graphically as is shown in Fig. 4.2.[3]

4.3 Early tools to measure Δt

The design of a practical tool to measure the time taken for a compressional wave to travel through 1 ft of formation was not easy. Even when an appropriate sound source was engineered and geophones arrayed to receive transmitted sound waves, the problem remained to avoid direct coupling from the transmitter to the receivers through the tool housing itself or through the mud column in the well being logged. Fig. 4.3 shows the original sonic logging tool design with one transmitter and two receivers.

This tool was built with the transmitter and the receivers mounted on a flexible "spine" suspended in a rubber tube filled with mineral oil. This

[3] In this chart, Δt_{fl} is assumed to be 189 μs/ft.

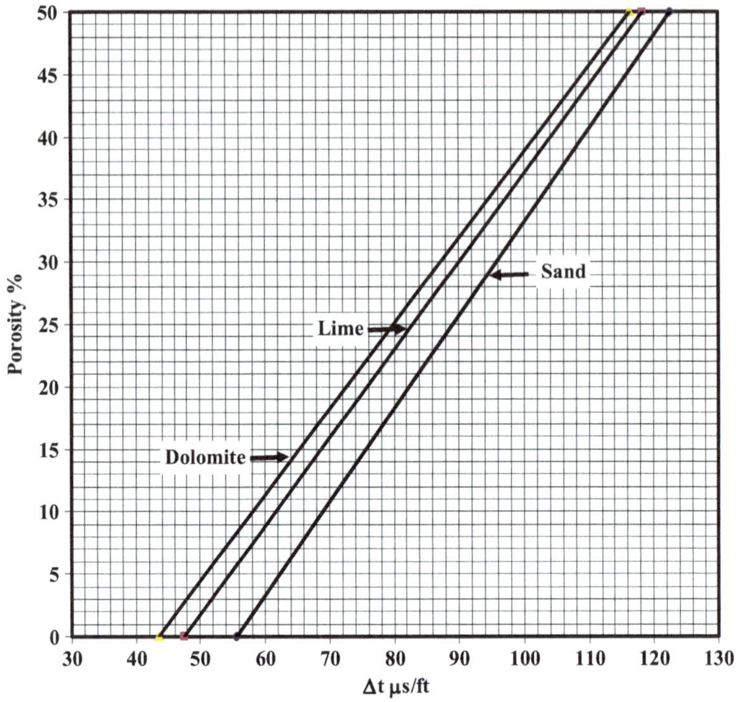

Figure 4.2 Porosity from sonic log for common rock types (Wyllie time average).

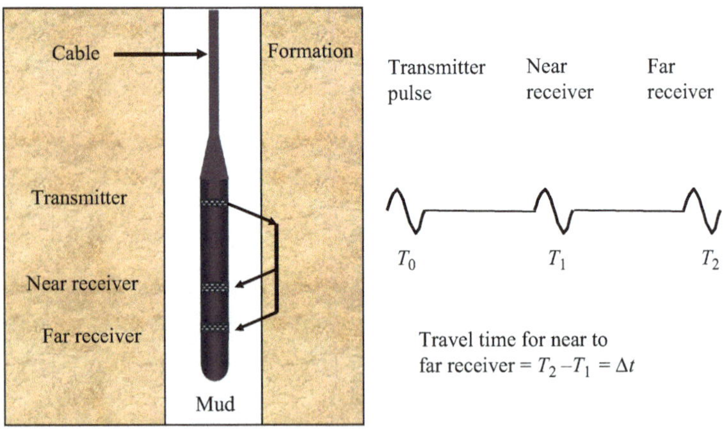

Figure 4.3 Early uncompensated sonic logging tool.

arrangement ensured that the time taken for sound to pass from the transmitter to the receivers through the formation was shorter than the time taken to go either through the mud in the wellbore or though the fluid-filled rubber sleeve itself. This presented some practical disadvantages in rugged or deviated

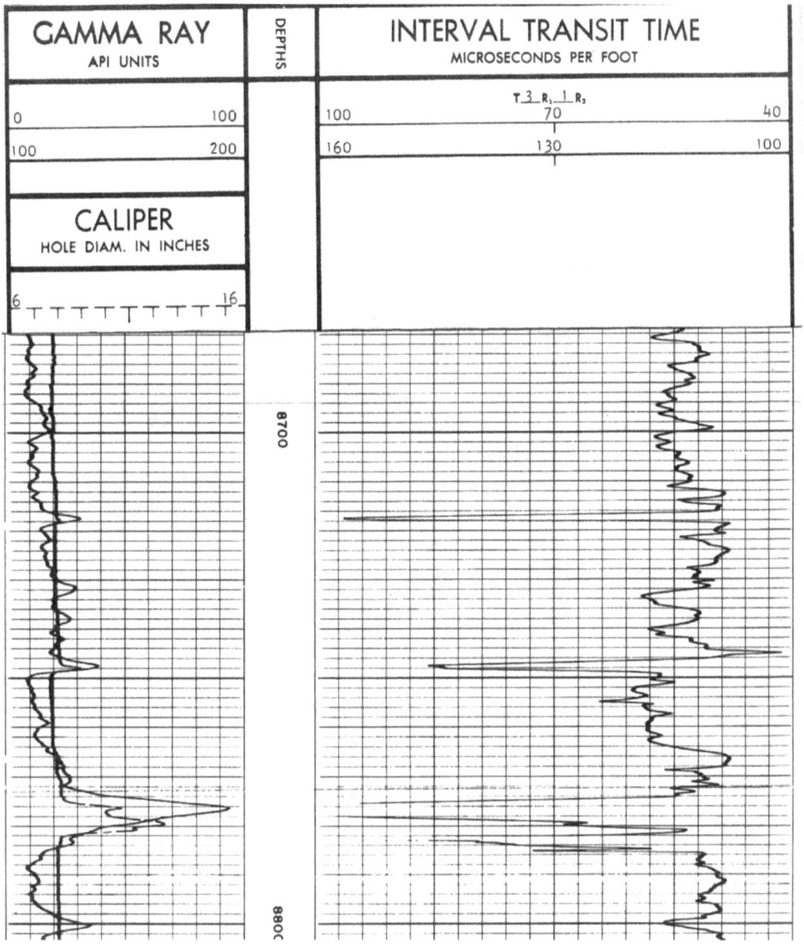

Figure 4.4 Uncompensated sonic log (warts and all).

holes and resulted in logs that were plagued with spurious "spikes" (cycle skipping) or distortions due to hole size changes. Fig. 4.4 illustrates an early uncompensated sonic log. "Spikes" are visible at 8716-18, 8746-49, and in the shale bed from 8774 to 8786. These are due to "cycle skipping" (false triggering on the first arrival of the sonic wave train) or to the uncompensated tool design being incapable of handling rapid hole size changes as evidenced from the caliper trace in the shale bed.

4.4 Borehole compensated sonic

Many of the deficiencies of the single transmitter/dual-detector sonic logging systems were corrected by the introduction of the borehole compensated

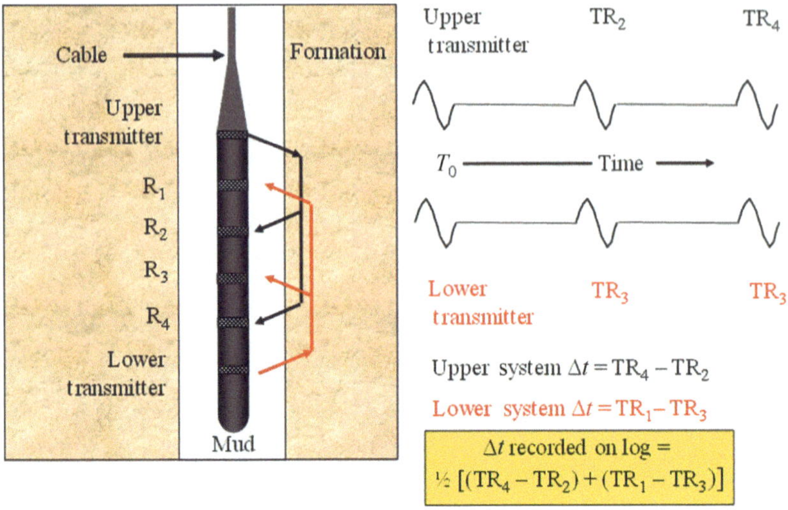

Figure 4.5 Borehole compensated sonic logging tool.

sonic tools in 1963. Fig. 4.5 illustrates the features of the compensation system employed by these improved devices.

The compensated tools used a specially machined rigid housing that avoided direct acoustical coupling from transmitter to receivers and effectively eliminated the "road noise" and cycle skipping problems of the original uncompensated tools. However, these tools could still only detect and measure the speed of compressional waves. Shear wave detection with sonic tools came after the end of the pre-digital age. Thus the uses of sonic logging tools in the pre-digital era were restricted to porosity, shale compaction trend monitoring, and lithology identification when used in combination with other measurements (e.g., neutron and/or density) and as an assist to the geophysicists running seismic surveys by means of the integrated travel time coupled with "check shots." Other more erudite uses were also found such as fracture indications and measures of the organic content of shales (total organic carbon [TOC]).

4.5 Sonic porosity transforms

As mentioned previously, Wyllie proposed a "time average equation" to relate the speed of sound to the porosity of a rock. Pickett also proposed and alternative relationship. These relationships gave good results in consolidated formations but failed in recent sediments such as Gulf Coast uncompacted shaly sands. For these formations, a modified version of the

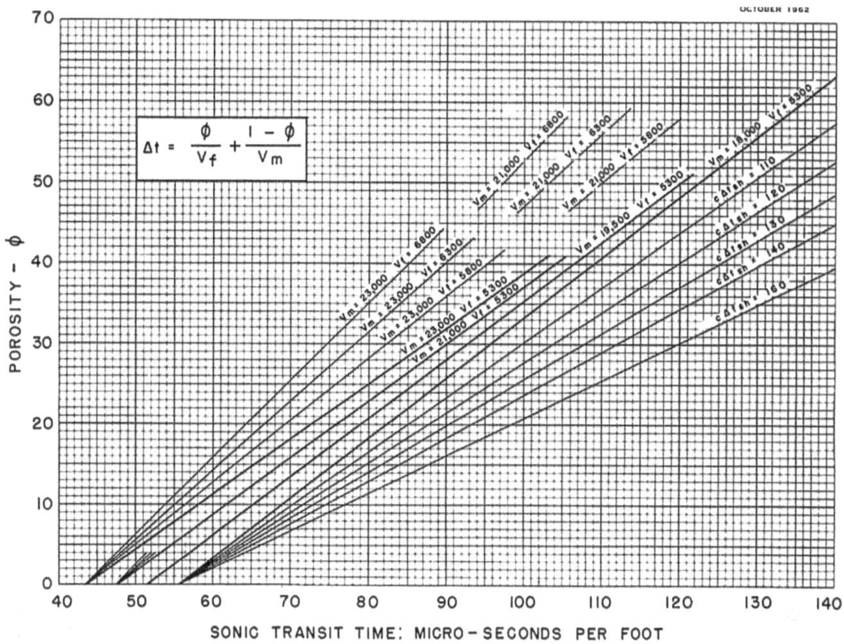

Figure 4.6 Sonic porosity chart for unconsolidated sediments.

Wylie plot was introduced as shown in Fig. 4.6. Note the "fan" of response lines emanating from the common starting point of 55.5 µseconds/ft, the travel time in sandstone. Each "spoke" of the fan is labeled with a value of "$C\Delta T_{sh}$." This parameter was a simple indicator of the degree of compaction of the sediments based on the travel time through shales surrounding the bed of interest. This factor could vary between 100 and 160. Thus a reading on a sonic log of 100 µseconds/ft in a sandstone could be interpreted as a porosity between 21% and 34% depending on the depth of burial and degree of compaction as indicated by the neighboring shales.

4.6 Overpressure detection

The dependence of the sonic travel time on both porosity and compaction was soon put to good use by a number of workers in the field who produced correlations between depth of burial, shale compaction, and pore pressure. Of particular mention in this respect is the work of William Lang whose definitive papers are referenced at the end of this chapter.

4.7 Sonic amplitude and fractures

Shortly after the introduction of the sonic logging tools that measure the speed of sound in the formation a secondary measurement came to be recorded, the amplitude of the compressional wave after it had traversed the formation from the transmitter to the receiver. It was found that the amplitude of the wave was attenuated in the presence of fractures. A number of well-site displays became popular in the 1960s and 1970s on which a sonic amplitude curve was displayed overlaying an amplified short normal resistivity curve. Depending on the formation logged and the scaling chosen for the two curves, the fractured section, if present, would be identified.

A similar approach was used in comparing a sonic travel time curve with a bulk density curve. The mutually compatible scales that worked best are shown in Fig. 4.7.

4.8 Use of sonic logs for lithology identification

When used in conjunction with other poro-lithology indicators, such as the neutron, density, and gamma ray logs, the sonic proved highly useful in the task of quickly and simply identifying rock types and minerals,

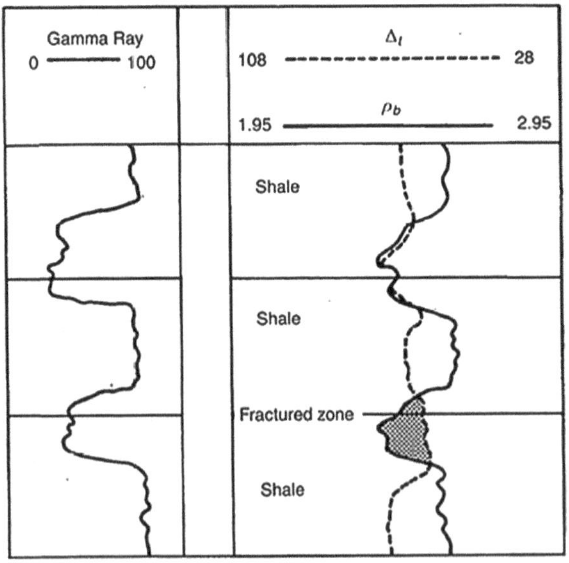

Figure 4.7 Specially scaled sonic and density traces for fracture identification.

where found in the logged columns. The materials in Chapter 15 provide details of the cross-plots and other methods used for this task and will not be repeated here.

4.9 Sonic-induction combination tool

The advent of combination tools opened the door to well-site log analysis and one of the first applications was the use of a sonic tool to provide a source of porosity that could then be combined with a measure of formation resistivity from an induction tool. The parameter chosen for a quick-look display was the apparent water resistivity given the symbol R_{wa}. The details of how this was arranged, and example logs are given in Chapter 13 and will not be repeated here.

4.10 Integrated sonic for seismic/check shots

Fig. 4.8 shows a section of a sonic log on which travel time integration is marked by a series of "pips." On Track 2, just to the right of the depth track can be seen marks extending one chart division to the right approximately every 10 ft. It can also be seen that after 10 of these small "pips," a larger one is registered extending across two chart divisions. The purpose of this display is to sum the travel time from any given depth in the well to any other depth or even to surface. Each small pip accounts for 1 ms and each big pip 10 ms. This type of sonic integration is useful to the geophysicist for comparison with any check shot that may have been recorded and helps in refining the all-important time–depth relationship that proper geophysical interpretation of a seismic survey requires.

4.11 Sonic measurements in organic shales

A simple and effective way of calculating TOC using a deep induction on a logarithmic scale overlaid with a scaled Δt trace from a sonic is familiar to today's analysts as the "Passey dlogR" method. Effectively, it is a close cousin to a Hingle plot, which uses a crossplot (rather than a scaled curve overlay) to highlight water and hydrocarbon bearing intervals using a porosity-related measure in the X-axis and the reciprocal root of R_t in the Y-axis. For a fuller treatment of the Hingle plot, the reader is referred to Chapter 11.

Although the Passey method has normally been used with post-digital well logs, there is no reason why the method cannot be used with

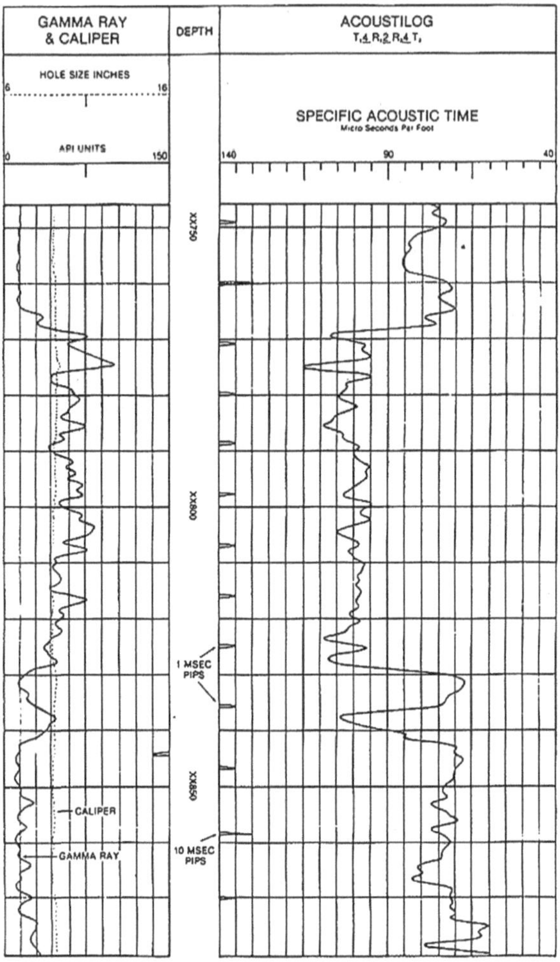

Figure 4.8 Integrated sonic log display.

pre-digital sonic and resistivity logs as well. All that is required is an overlay of the sonic on a linear scale with the induction on a logarithmic so that two decades of deep resistivity coincide with 100 μseconds/ft of Δt. Fig. 4.9 gives an example of such an overlay.

The mathematical expression for calculating dLogR is given by:

$$\text{dLogR} = \log_{10}\left(R/R_{baseline}\right) + 0.02(\Delta t - \Delta t_{baseline})$$

In turn, the dLogR number is related to TOC via a previous knowledge of the "level of maturity" (LOM). Fig. 4.10 offers an overview of the dLogR plot in formations of varying levels of maturity. The Passey

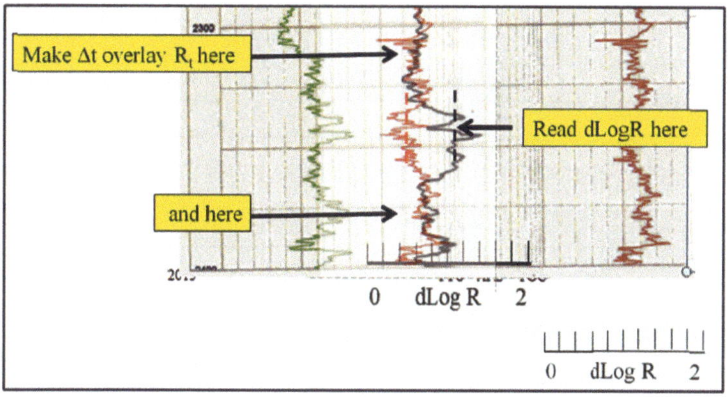

Figure 4.9 Resistivity and sonic overlay for Passey method derivation of total organic carbon.

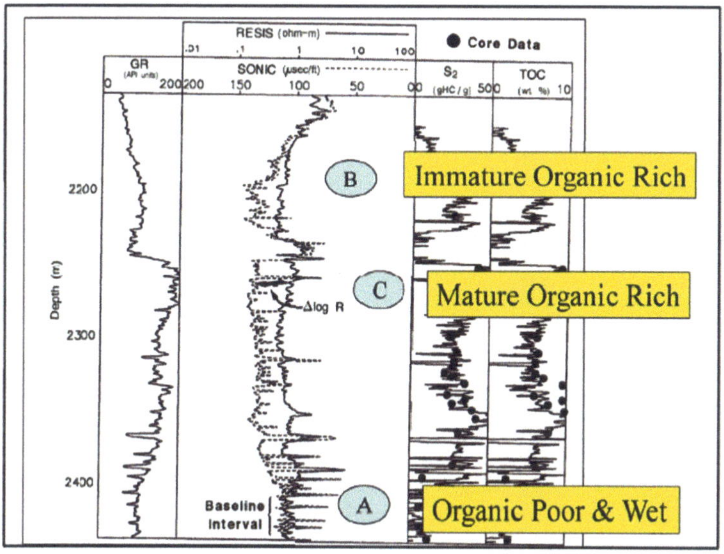

Figure 4.10 dLogR plot for different organic richness. Courtesy Passey et al.

method for TOC can be appreciated by the plot shown in Fig. 4.11 that contrast it with the Hingle method. Essentially both use the same interplay of resistivity and sonic to highlight where there are hydrocarbons and where there is water.

Once the dLogR has been evaluated, the actual TOC is derived from:

$$\mathrm{TOC} = (\mathrm{dLogR})x10^{(2.3-\mathrm{LOM}/6)}$$

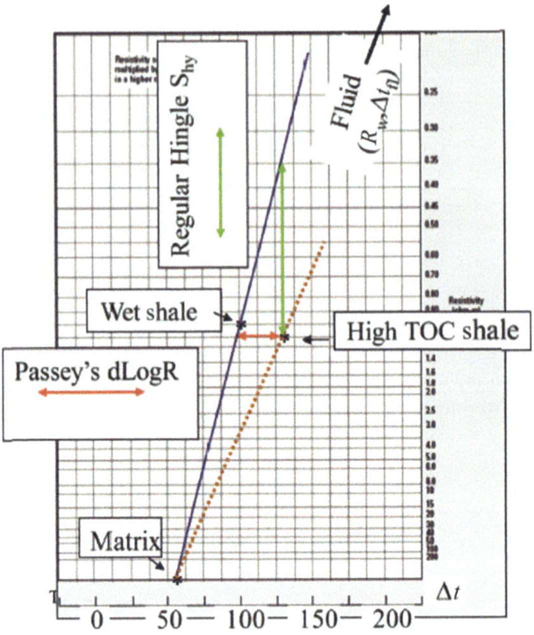

Figure 4.11 Comparison of Hingle plot to Passey method.

The value of LOM may not be available from vitrinite reflectance if one is working with pre-digital logs, in which case local, "tribal" knowledge can be useful. Failing that the Zhao Method[4] can be used that suggests that:

$$\text{LOM} = \frac{1}{\{\phi_N x\sqrt{(1 - S_w)}\}}$$

Other methods include the Suggate's coal ranking scale (from 1959) and/or the work of Bostick and Damberger (1971). At a last resort, there are always the Schmoker Correlations. The relative one here for use with a sonic log would be:

$$\text{TOC} = \Delta t\ x\ 0.343 - 36$$

In summary, the (pre-digital) measurements of the speed and amplitude of compressional waves traveling through porous media added enormously to the analyst's understanding of the formations through which they passed. Building on these early tools and analysis methods, the

[4] See AAPG Bulletin, Vol. 91, No. 4 (April 2007), pp 535-549

modern advances afford even deeper understanding of the more complex reservoir challenges faced today.

Further reading

Anderson, T., Walker, T., 1972. Log derived rock properties for use in well stimulation design. 47th Annual Fall meeting of the Society of Petroleum Engineers of AIME. 8–11 October, San Antonio, Texas.

Anderson, W.L., Walker, T., 1961. Application of open hole acoustic amplitude measurements. Paper SPE-122, 36th Annual Fall Meeting. SPE of AIME, Dallas.

Biot, M.A., 1952. Propagation of elastic waves in a cylindrical bore containing a fluid. J. Appl. Phys. 23 (9).

Kean, C.H., Tulles, F.N., 1948. Acoustic impedance logging. Oil Gas J. 47, 95.

Lang, W.H., 1978. The determination of prior depth of burial (uplift and erosion) using interval transit time. Paper B in Annual Well Logging Symposium Transactions. Society of Professional Well Log Analysts, pp. 13–16.

Lang, W.H., 1994. The determination of thermal maturity in potential source rocks using interval transit time/interval velocity. Log Analyst 35 (6).

Millican, M.L., 1960. The sonic log and the delaware sand. J. Pet. Technol. 12, 71.

Morris, R.L., Grine, D.R., Arkfeld, T.E., 1963. The use of compressional and shear acoustic amplitudes for the location of fractures. SPE Paper No. 723, 38th Annual Fall Meeting of the Society of Petroleum Engineers of the American Institute of Mining, Metallurgy and Petroleum Engineers. 6–9 October.

Passey, Q.R., Creaney, S., Kulla, J.B., et al., 1990. A practical model for organic richness from porosity and resistivity logs. Am. Assoc. Pet. Geol. Bull. 74 (12), 1777–1794.

Pickett, G.R., 1963. Acoustic character logs and their applications in formation evaluation. Jour. Pet. Tech. 15 (6), 660–667.

Strick, E., 1959. Propagation of elastic wave motion from an impulsive source along a fluid/solid interface, II Theoretical pressure response, III The pseudo-rayleigh wave. Phil Trans. Roy Soc. London, Ser. A 251 (1000).

Summers, G.C., Broding, R.A., 1952. Continuous velocity logging. Geophysics 17 (3), 435–649.

Tixier, M.P., Alger, R.P., Doh, C.A., 1959. Sonic logging. Jour. Pet. Tech 1 (5).

Tixier, M.P., McVicar B.M., Burton, R.P., 1960. Progress in sonic log application. The First Joint Tech. Meeting of the CIM Petroleum and Natural Gas Div. and Rocky Mountain Sections of SPE of AIME. 4–6 May, Calgary, Alberta.

Walker, T., 1962. Progress report on acoustic amplitude logging for formation evaluation. Paper SPE-451, 37th Annual Fall Meeting. SPE of AIME, Los Angeles.

Walker, T., Riddle, G., 1963. Field investigation of full wave acoustic wave recording. Paper for Fourth Annual Logging Symposium. Society of Professional Well Log Analysts, Oklahoma City.

Wyllie, M.R.J., Gardner, G.H.F., Gregory, A.R., 1962. Studies of elastic waves attenuation in porous media. Geophysics 27 (5), 569–724.

CHAPTER 5

Neutron logging and interpretation

Abstract

At the time the first electric log was completed, the neutron was still yet to be discovered. James Chadwick conducted his experiments that proved the existence of neutrons, as distinct neutral particles, in 1932. He showed that the neutron had the same mass as the proton but with much greater penetrating power. (He was awarded the Nobel Prize for his discovery in 1935.) It took only 16 years from the discovery of the neutron to the introduction of neutron logging to the formation evaluation arsenal. This chapter will chronical and explain how these minute particles have led to quantum leaps in formation evaluation.

Reproduced from "Views of Schlumberger" by Norman James with kind permission of the author

Formation Evaluation with Pre-Digital Well Logs
DOI: https://doi.org/10.1016/B978-0-12-820232-6.00005-4

5.1 Origins of neutron logs

Neutron logging was introduced in 1948. The first tools used a chemical source of neutrons and had a single detector. The log recorded the "Neutron count rate" in *Counts per second*. It was up to the analyst to calibrate these raw measurements into meaningful values for porosity or to determine the hydrocarbon type (oil or gas) if present.

These devices responded to anything that contained hydrogen. This hydrogen dependence can be seen as both a blessing and a complication. The primary response is to porosity since the pore space will contain hydrogen in the form of water and/or hydrocarbons. However, some of the matrix materials may also have a significant hydrogen content—such as in the case of the clay minerals found in shales.

Chemical neutron sources are intimate mixtures of Beryllium and other elements that are alpha emitters, such as radium, plutonium, or americium. Such sources are characterized by how many neutrons per second they emit, which in turn depends on the mix of alpha emitters used. At first, the logging companies each had their own system of units on their recorded logs. Schlumberger used "cps," PGAC used "standard neuron units," and Land Wells "environmental units." The American Petroleum Institute (API) brought some order to the scene with an "API Neutron Unit." This was defined with reference to a special calibration pit at the University of Houston where the difference between the reading with and without the source was defined as 1000 API units.

The detectors used in the early tools did not in fact detect neutrons. Rather, neutrons reaching the detector would find a regular "Geiger" detector of gamma rays wrapped in a cadmium shield. This would absorb the incident neutrons by capture and emit γ-rays that were then counted in the Geiger counter.

5.2 Life cycles of neutrons

On leaving the source a neutron is considered "Fast." As it traverses the tool housing, the mud in the borehole and the formation it gets slowed down by multiple interactions with the nuclei in its path. From the initial *Fast* stage, it will pass through an intermediate *Epithermal* stage to end up at a *Thermal* stage, when it becomes ripe for capture by the nucleus of an element in its path. Such a capture is accompanied by the emission of a gamma ray. Along this chain of events, illustrated in Fig. 5.1, a number of different

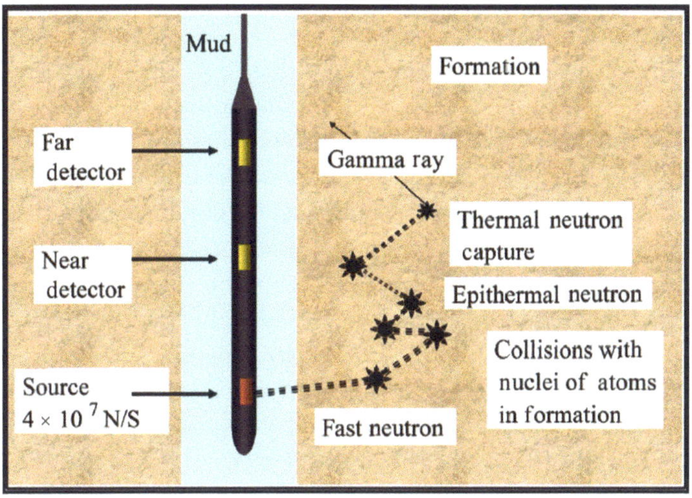

Figure 5.1 the Life cycle of a fast neutron.

logging tools have been built and brought into good use. These include the early "count-rate neutrons," the epithermal or "sidewall" (SNP-type) neutron tools, and the compensated thermal neutron (CNL-type) tools. Note that although the first tools had a single detector, the illustration in Fig. 5.1 includes a second detector since the subsequent thermal neutron tools used two detector as will be discussed later (see Chapter 7).

5.3 Neutron tool chronology

Starting in the late 1940s, neutron logging tools have progressed in step with technology. Basically there are three generations of neutron porosity tools that will be found in any collection of pre-digital paper log files. These were:

- single detector of neutrons (unspecified), uncalibrated (count-rates, API units, etc.), centered in borehole tools,
- single detector of epithermal neutrons, calibrated in %, eccentered (sidewall pad), and
- dual detector of thermal neutrons, calibrated in %, semi-eccentered (bow spring assisted).

For those tools in use in the pre-digital era, their idiosyncrasies will be discussed along with guidelines for calibration and interpretation.

5.4 Count-rate neutron tools

A typical early neutron log is shown in Fig. 5.2. The neutron curve is shown in the far-right track. The track on the far left is an SP, and the two tracks in the middle are resistivity tracks. Note that low counts correspond to high porosity and high counts to low porosity. Common to the myriad of interpretation/calibration charts published by the logging companies the basic saving grace is that for these early "count-rate" tools, the logarithm of the porosity is linear with the count rate. Thus the analyst only needs to set up a local calibration chart to find the needed answer for porosity. Fig. 5.3 illustrates the principal of the method. Two calibration points are needed.

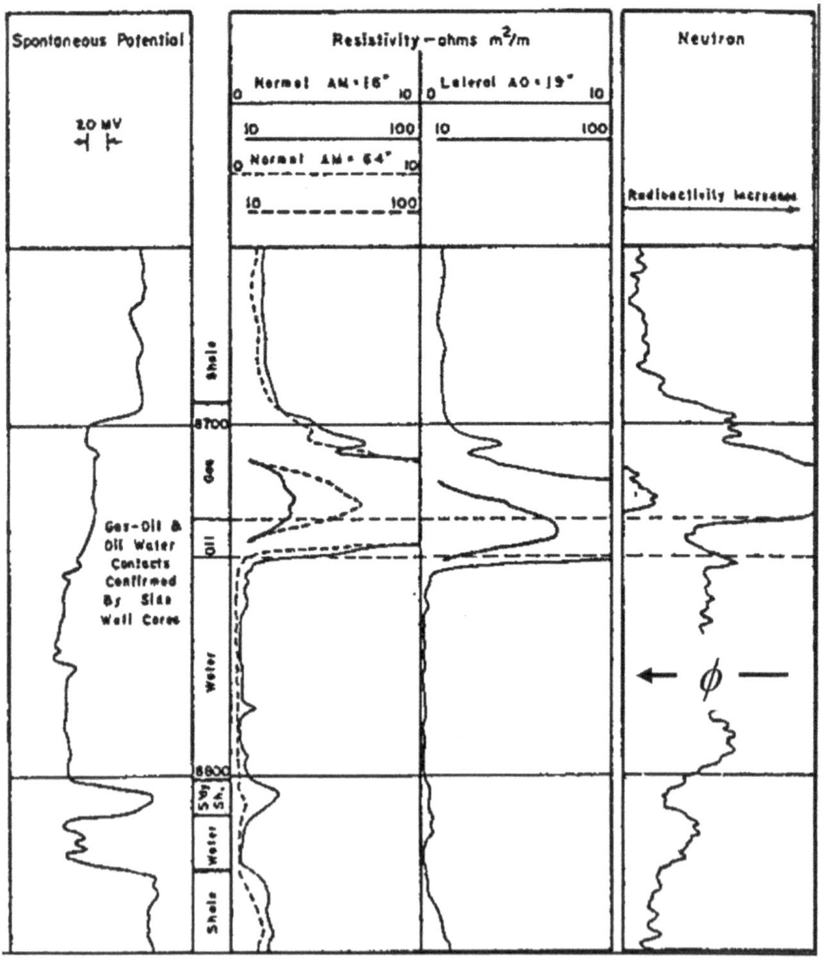

Figure 5.2 Early "count-rate" neutron log.

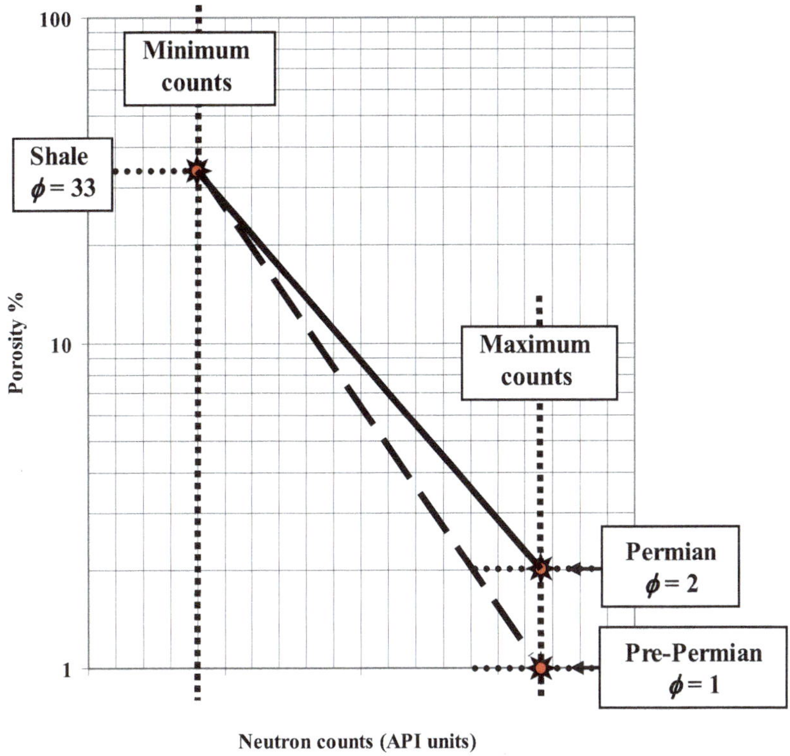

Figure 5.3 Bootstrap method for neutron calibration.

The first will be where the neutron count-rate (be it in "cps" or API units) is lowest, which will usually be where the tool has logged past a shale section. This can be independently confirmed by observation of the SP and/or gamma ray response. The second calibration point will be where the count-rate is highest, which will be in a consolidated low porosity carbonate formation. These two points can then be posted on a sheet of log-linear graph paper with the log scale (two decades) on the *Y*-axis and the linear scale, as convenient, on the *X*-axis. As a starting point, the count-rate in the shale should be posted at a porosity of 33% and the count-rate in the tight formation at 2%, if the calibration formation age is Permian, but 1% is pre-Permian.

Subsequently the porosity at any chosen point may be easily read by entering the calibration chart (see Fig. 5.4) with the "counts" and moving up to the calibration line to read the porosity on the *Y*-axis. Depending on the logging company, and the year the log was run, the *X*-axis may have to be scaled in counts, cps, API neutron units or whatever the scaling may have been for that time and place.

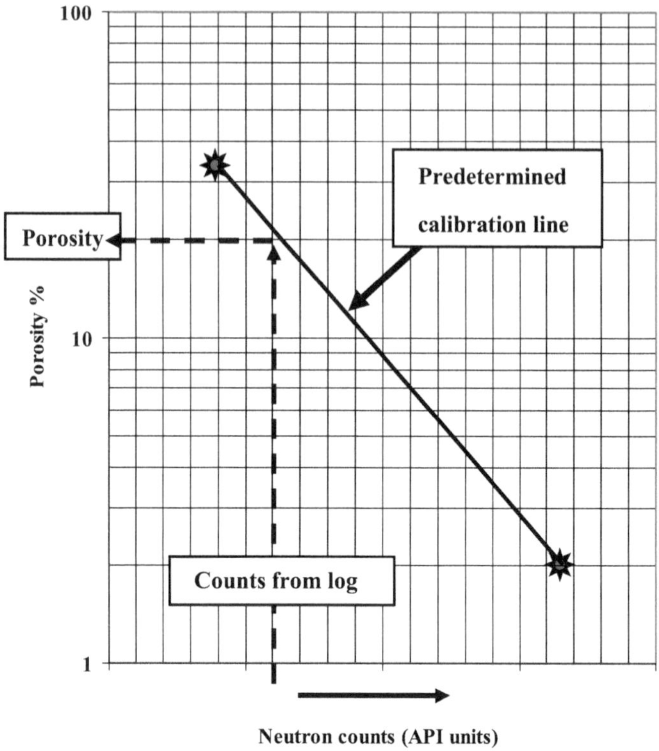

Figure 5.4 Reading porosity from neutron count-rate calibration line.

5.5 Dual-spaced neutron for gas detection

In many cases, the neutron tool was run twice over the same interval. Once with the source to detector spacing set to its normal position and then on a second trip into the hole but with a spacer introduced between the source and the detector. The two traces so obtained were them superimposed to reveal any gas-bearing intervals in the logged section. Fig. 5.5 illustrates this technique.

Note the interval from 4525 to 50 where the dashed curve separates from the continuous curve to highlight the presence of lighter hydrocarbons. Effectively the longer spacing allows the longer spaced detector to monitor neutron interactions deeper into the formation away from the flushed zone immediately adjacent the wellbore. Having a lower density of hydrogen atoms, due to gas being less dense than liquid hydrocarbon, is, for the neutron tool, the same as having a lower "porosity" and hence the higher count-rate in the gas-bearing intervals.

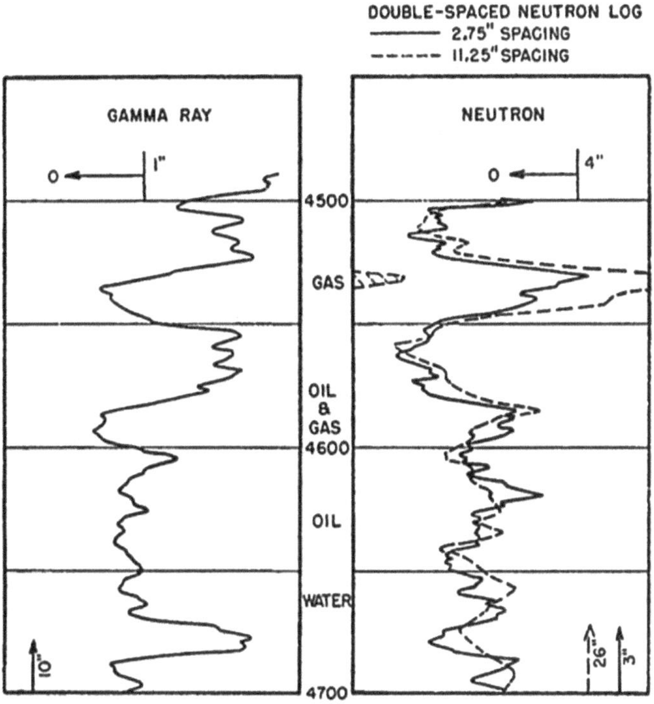

Figure 5.5 Double-spaced neutron log for gas detection.

5.6 Neutron logging in cased hole

Very early on the usefulness of the neutron logging tool was not lost on the industry since it could be run both in open hole and in cased hole since the neutrons and the gamma rays could penetrate steel casing. Fig. 5.6 shows a neutron log run in both the open hole and the cased hole. Note the scales for "Run #1" is 50 "Standard Neutron Units" per track and for "Run #2" 30. Although the two traces have almost identical character, the cased-hole trace has fewer "units" at any given point in the well and also the entire trace is reduced by a factor of about 25% due to the radiation absorption by the casing and cement.

Only slightly later in the development of neutron logging, the Dresser Atlas company fielded a combination tool that included a neutron curve, obtained from a conventional epithermal sidewall neutron pad mounted on a mechanical carrier equipped with a hydraulic retractor and a sidewall acoustic section mounted on the same carrier but contacting the opposite side of the borehole is. The entire acoustic section, consisting of a transmitting transducer and two receiving transducers, was

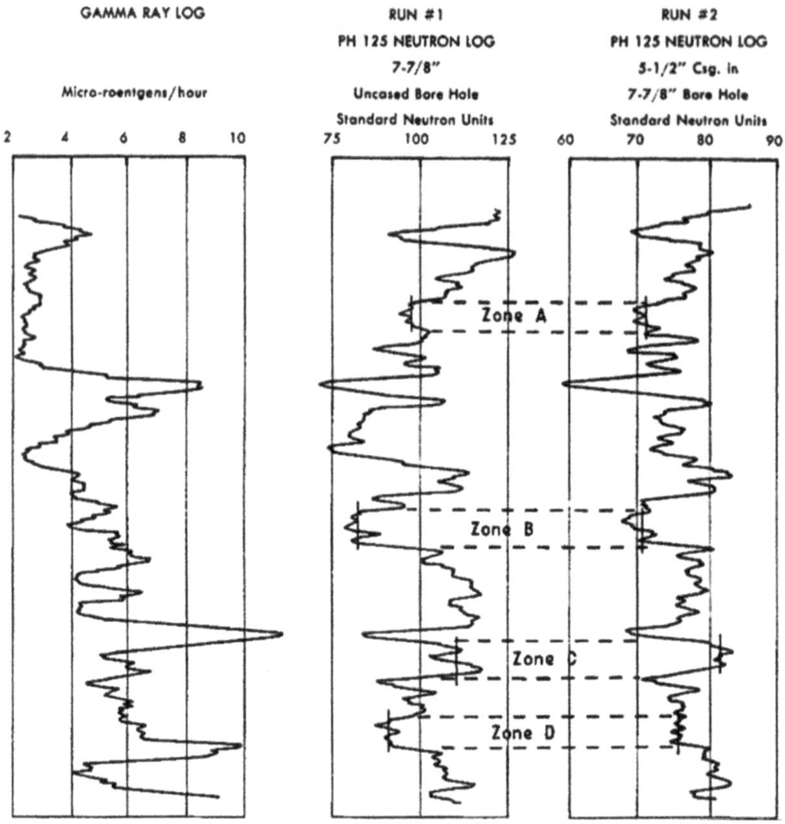

Figure 5.6 Neutron log in open hole and cased hole. Recorded by PGAC (date unknown).

contained in a pad approximately 2 ft in length. The receiver spacing was 6 in. producing an acoustic interval transit time curve (Δt) with much sharper interface resolution than that obtained from a conventional acoustic logging device.

5.7 Neutron log response to gas

In analyzing any neutron log, there are factors be taken into account. They include the radial saturation profile, the neutron logging tool's radial response profile, and the hydrogen index of the hydrocarbons present in the volume of rock sensed by the tool. Fig. 5.7 gives a generalized schematic of a generic neutron logging tool and a view of the annular volumes of the logged formation.

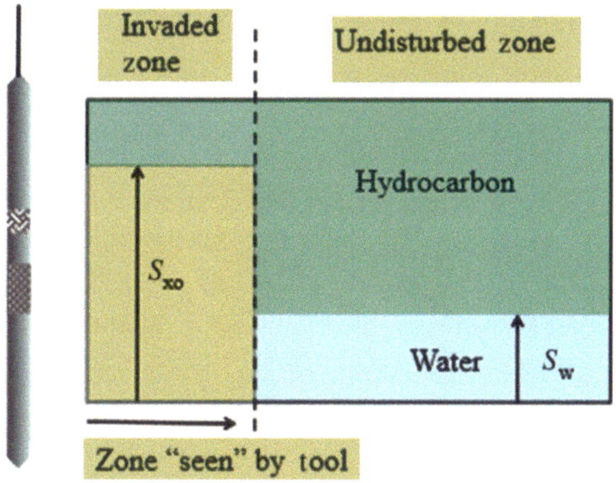

Figure 5.7 Neutron tool's view of mud filtrate, connate water, and hydrocarbon.

The neutron response is thus dependent on what lies within the volume that the tool senses. In general, the count-rate neutron tools, the sidewall neutron tool(s), and the formation density tool(s) get about 50% of their total signal from the first 2 in. radially outward from the borehole wall and 90% from the first 4 in.[1]. This means to say that in most cases, these tool make their reading in the flushed (invaded) zone where the water saturation is S_{xo}. Thus, what is seen on the log is given by:

$$\phi_N = \phi_{Nma}(1-\phi) + \phi_{Nmf}\phi S_{xo} + \phi_{Nhy}\phi(1-S_{xo})$$

Where

ϕ_{Nma} is the response to the matrix (usually considered zero)
ϕ_{Nmf} is the response to mud filtrate (usually considered equal to one)
ϕ_{Nhy} is the response to hydrocarbon

The neutron response to mud filtrate is mildly dependent on the salinity and varies between 1.0 for fresh filtrates and 0.9 for salt-saturated filtrates. However, the response for varying hydrocarbon densities is more marked. Fig. 5.8 shows this dependency.

ϕ_{Nhy}, the neutron response to hydrocarbon is related to the actual hydrocarbon density by:

$$\phi_{Nhy} = 2.2\rho_{hy} - 1.2\left(\rho_{hy}\right)^2$$

[1] The dual-spaced compensated neutron tool(s) read deeper into the formation with 50% of their signal coming from the first 4″and 90% from the first 10″.

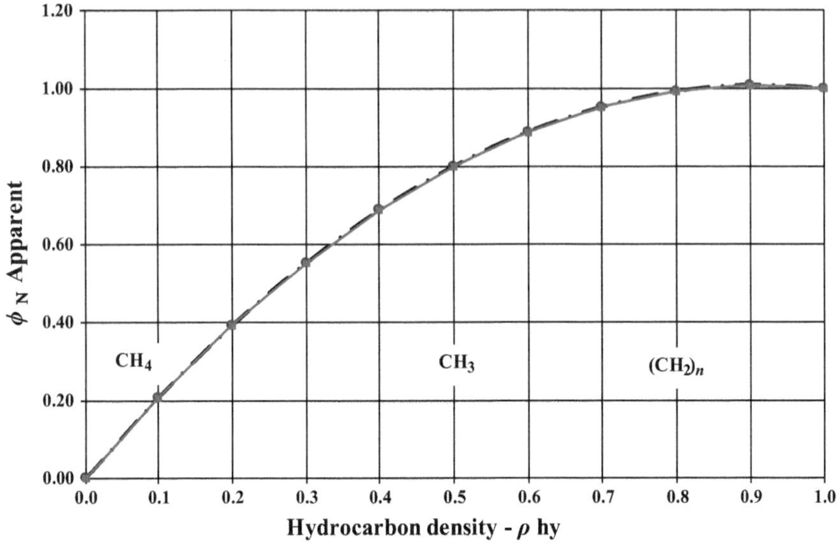

Figure 5.8 ϕ_{Nhy} as a function of hydrocarbon density in g/cc.

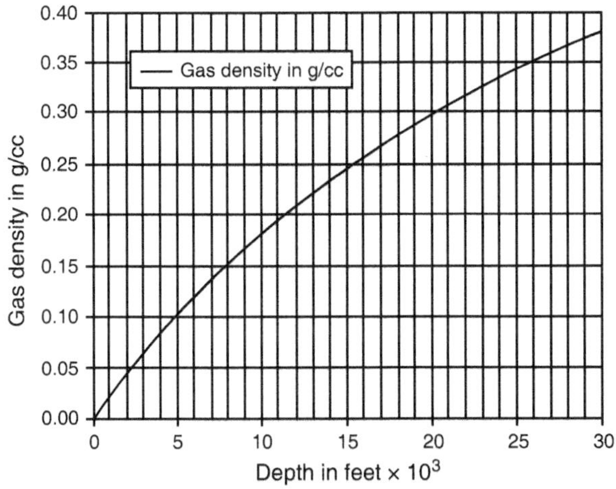

Figure 5.9 Gas density as a function of depth.

This is shown on the chart of Fig. 5.8. It is evident that when gas is present in the formation, the neutron tools record apparent porosities that are smaller than true porosities. The magnitude of the apparent porosity reduction may be estimated by a simple exercise. Fig. 5.9 shows the average methane gas density as a function of depth of burial for normally pressured formations with an average thermal gradient.

By way of example, let us suppose that we come across a neutron log recorded in a formation at 8000′ depth. From Fig. 5.9, we can read that the gas density at that depth will be 0.15 g/cc. From Fig. 5.8, we see that a hydrocarbon gas with a density of 0.15 g/cc will appear to a neutron tool as having a response of 0.3. The full response of the neutron logging tool (what we see on the log) with therefore be given by:

$$\phi_N = \phi S_{xo} + 0.3\phi(1 - S_{xo})$$

We may not know the true value of S_{xo} until after the porosity has been determined and an evaluation made by Archie's equation using measured values of R_{xo} and R_{mf}. However, the purpose here is to demonstrate that the apparent porosity from the neutron tool is less than the true porosity when gas is present in the rock volume investigated. Simple substitution in the equation above assuming the S_{xo} value to be 75% yields:

$$\frac{\phi_N}{\phi} = S_{xo}(1 - 0.3) + 0.3 = 0.75\ x\ 0.7 + 0.3$$

which evaluates to 0.825. So, if the true formation porosity is 30%, the neutron tool will say it is 24.75%.

Further reading

Amaldi, E., D'Agostino, O., Fermi, E., Pontecorvo, B., Rasetti, F., Segre, E., 1935. Artificial radioactivity produced by neutron bombardment. Proc. R. Soc. London, Series A 149 (868), 522-55.

Anon, 1940. Discussions and communications; radioactive well logging service. Geophysics 5 (4), 402.

Bailey, B.H., Bryant, H.L., Powell, N.L., 1957. Spot those gas zones with neutron logging. Oil Gas J. 368–370.

Belnap, W.B., Dewan, J.T., Kirkpatrick, C.V., Mott, W.E., Pearson, A.J., Rabson, W.R., 1945. Api calibration facility for nuclear logs. Published by the API in November as Recommended Practice RP33.

Blanchard, A., Dewan, J.T., 1953. The calibration of gamma ray logs and experimental basis for neutron logging interpretation. Pet. Eng. 25 (8), B76, B78-80.

Brown, A.A., Bowers, B., 1958. Porosity determination from neutron logs. Pet. Eng. B 30–34.

Dewan, J.T., 1962. Neutron log correction charts for borehole conditions and bed thickness. J. Pet. Technol. 8, 50–58.

Dewan, J.T., Allaud, L.A., 1953. Experimental basis for neutron logging interpretation. Pet. Eng. 25, 5 pp.

Pontecorvo, B., 1941. Neutron well logging. Oil Gas J. 40 (18).

Recommended Practice for Standard Calibration and Form for Nuclear Logs, 1959. API RP 33. American Petroleum Institute, New York.

CHAPTER 6

Gamma – gamma density logs

Abstract

The term "gamma – gamma" refers to the first formation density-logging tools that used a chemical source of gamma (γ) rays and a single detector of γ-rays—after they had been scattered in the formation logged. Later, improved, versions of the density-logging technology employed two γ-ray detectors that allowed for some compensation for the presence of mud cake. These latter tools were referred to as compensated density tools and are familiar to today's analysts. However, the original tools did not make their recordings directly in units of density (g/cc), as do the tools used today, but in a variety of units that require conversions to something useful and more directly related to formation porosity. These conversions are the main topic of this chapter.

Parkersburg semi-portable drilling machine. *Courtesy Parkersburg Rig and Reel Company.*

DOI: https://doi.org/10.1016/B978-0-12-820232-6.00006-6

6.1 Early density-logging summary

Introduced in 1956, the formation density log used a gamma ray source and a single gamma ray detector, both mounted on an articulated "pad" that was forced into contact with the wall of the wellbore. The logs were recorded in units of "counts per second" (cps) and needed a special calibration to go from cps to density in g/cc. These tools provided an interpretation path to formation porosity but behaved poorly in rugose holes and had no way to compensate for mud cake that could come between the pad and the formation. Later on the compensated density tool was introduced with a single gamma ray source but with *two* gamma ray detectors, which provided some degree of compensation for mud cake. The newer tools recorded formation density directly in units of g/cc, which made interpretation much easier.

6.2 The uncompensated density tool

Fig. 6.1 shows a schematic of the original, single source/single detector, tool. Charts were available to convert from "standard" counts per second to formation bulk density (in g/cc) as shown in Figs. 6.2 and 6.3.

These charts were also available as a single chart (see Fig. 6.4) that handled both gas and liquid-filled holes and, in addition, allowed the analyst to convert the formation bulk density, read from the log, into a porosity value depending on the matrix material present in the formation (sand, lime, and dolomite, as shown in the below figure).

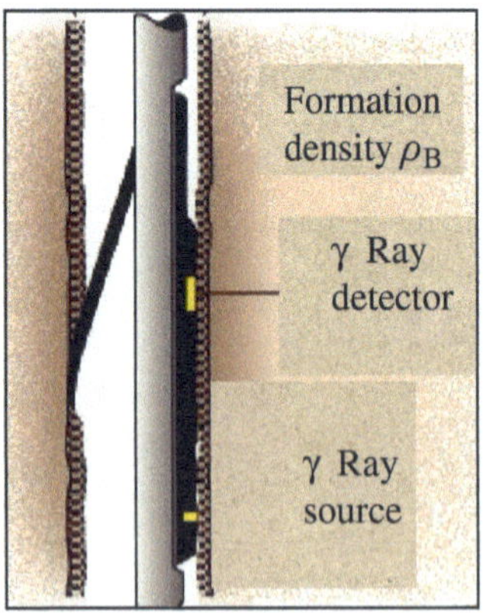

Figure 6.1 Single detector "γγ" density tool.

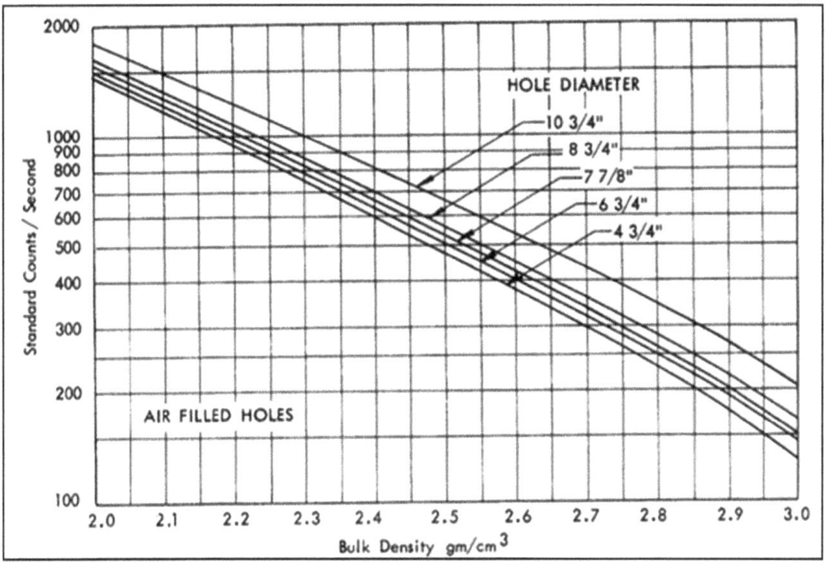

Figure 6.2 γγ counts per second to bulk density (air-filled holes). *Image Courtesy Schlumberger.*

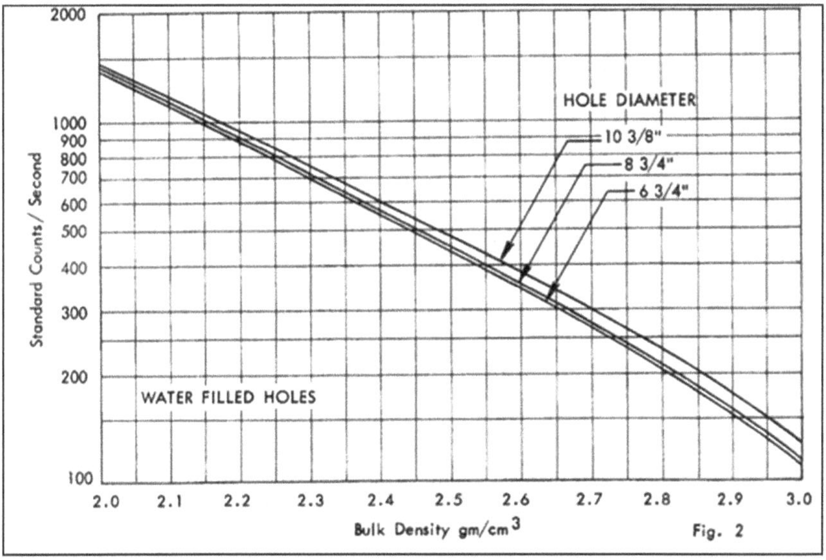

Figure 6.3 γγ counts per second to bulk density (liquid-filled holes). *Image Courtesy Schlumberger.*

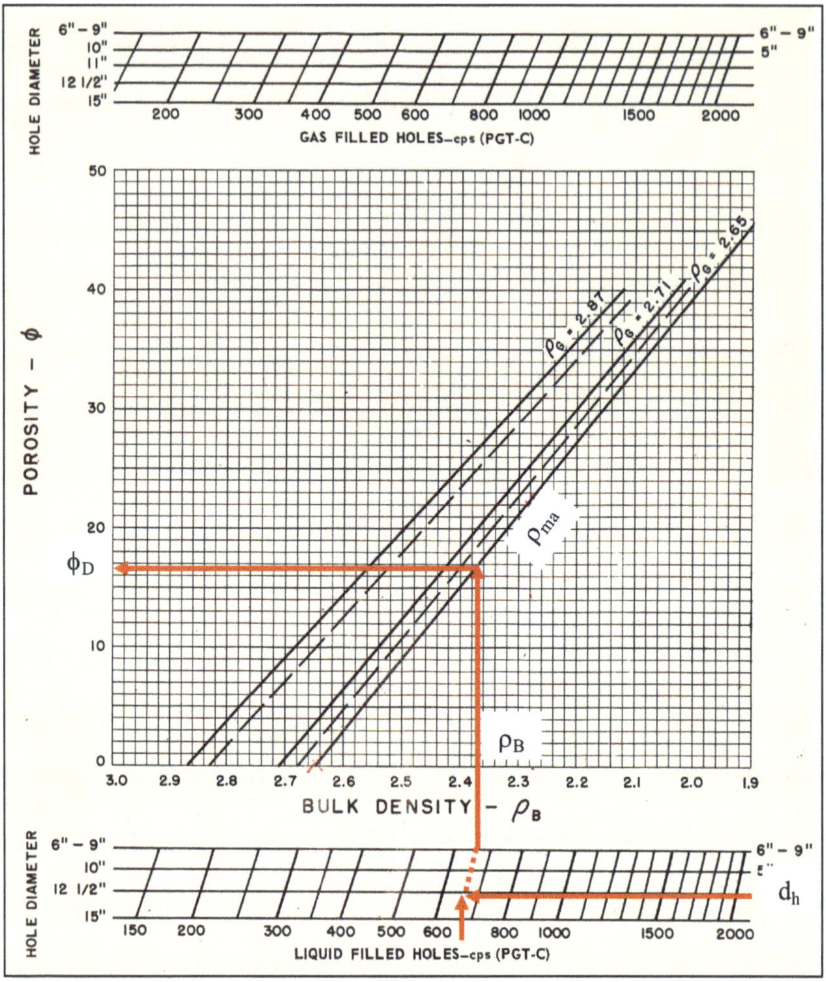

Figure 6.4 γγ counts to porosity. *Image Courtesy Schlumberger.*

The example shown on Fig. 6.4 is for a log-recorded 675 cps in a 12½″ diameter, liquid-filled hole. The interpreted bulk density is 2.365 g/cc, which leads to the sandstone porosity of 17%. Some of the early density logs that were recorded in cps would label them as standard density units (SDU), and the log headers would show a conversion to grain density-dependent porosity. An example is shown in Fig. 6.5. A log value of 1200 SDU (cps) corresponds to a bulk density of 2.4, which in turn translates into a sandstone porosity of 15%.

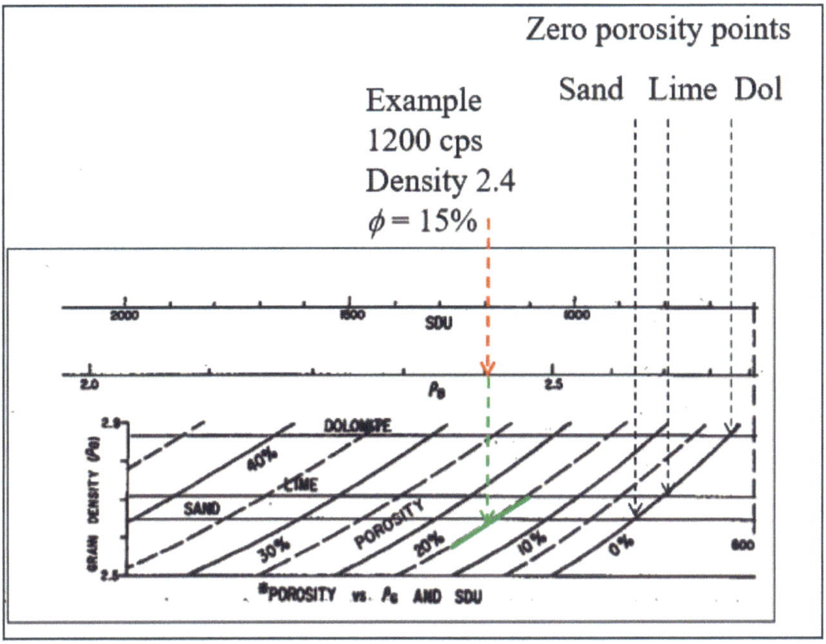

Figure 6.5 Early density log in SDU with porosity scaling on header.

6.3 The compensated density tool

The underlying physics of a bulk density measurement is a log-linear relationship between counts per second and bulk density. It has the form:

$$\rho_B = A - B \log(\text{counts/second})$$

This general relationship holds for not only the single detector ($\gamma\gamma$ density tool) but also the compensated dual detector tool that was to follow. Fig. 6.6 illustrates the compensated density tool with the dual detectors.

A linear response line may be drawn to mark off the actual formation density by plotting the count rates from the two detectors. Points that fall off the trend line reflect the insertion of mud cake between the tool face and the formation. The mud cake, having a density different from that of the formation (either higher or lower), moves the plotted point away from the mud cake-free "spine" along a "rib." This is shown in Fig. 6.7.

The case illustrated is for a formation with bulk density of 2.7 g/cc and a mud cake with a density of 1.5 g/cc. The dotted line represents all possible cases between zero mud cake thickness (t_{mc}) and a case of infinitely thick mud cake. The compensated tools thus cleverly took care of providing the analyst with true formation density values from which to calculate porosity.

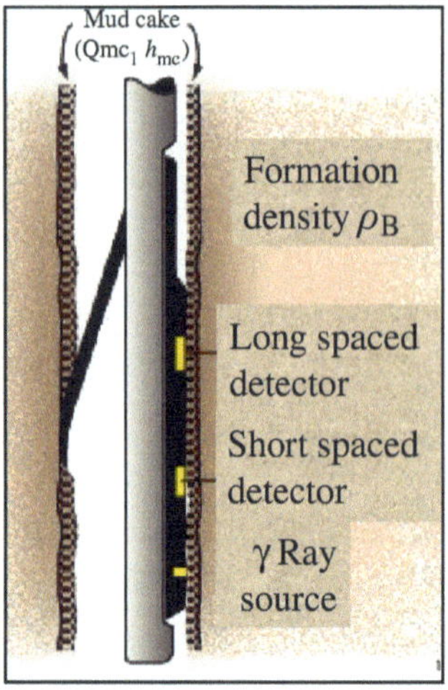

Figure 6.6 Dual detector formation density tool.

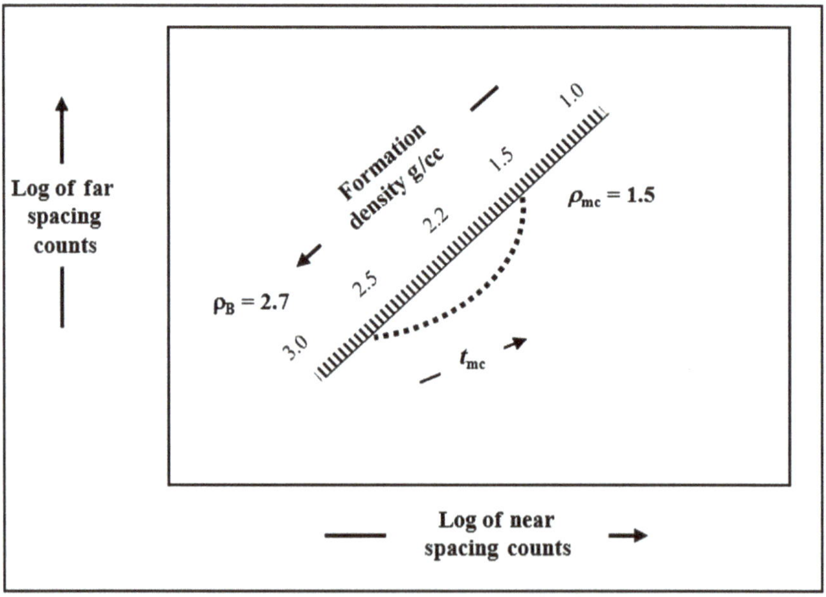

Figure 6.7 Compensated density tool near and far counts.

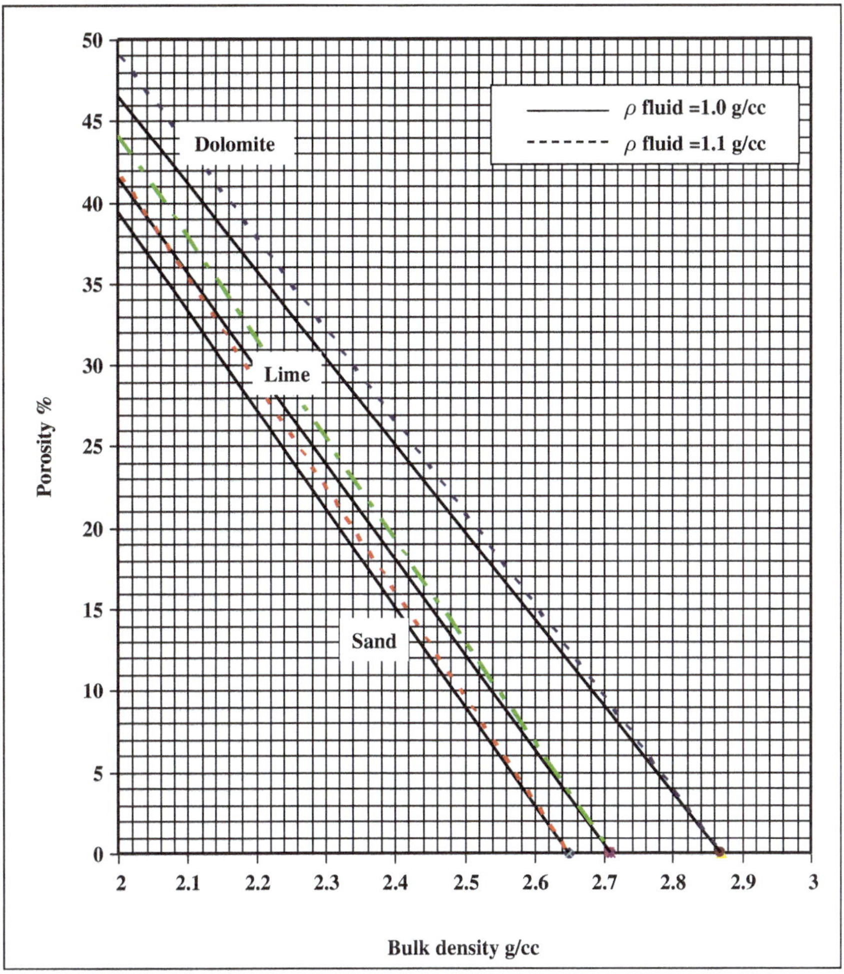

Figure 6.8 Porosity from bulk density for common matrices and mud filtrate densities.

6.4 Porosity from formation density

Fig. 6.8 is a graphical solution to the simple equation that relates bulk density to porosity.

$$\rho_B = \phi\rho_{Fluid} + (1 - \phi)\rho_{ma}$$

From which is derived:

$$\phi = \frac{\rho_{ma} - \rho_B}{\rho_{ma} - \rho_f}$$

The fluid in the pore space measured by the density tool will be a mix of mud filtrate, connate water, and residual hydrocarbons. It is customary to assume that the fluid is just mud filtrate that may have a density between 1.0 and 1.1 g/cc, depending on the salinity of the mud system used to drill the hole. In the case that there are residual hydrocarbons as well, then the fluid density will be controlled by both the type of hydrocarbon (oil or gas) and its saturation:

$$\rho_f = \rho_{mf} S_{xo} + (1 - S_{xo})\rho_{hy}$$

The chart of Fig. 6.8 is just built for fresh and salty mud filtrates and the most commonly encountered matrix materials—sandstone, limestone, and dolomite.

6.5 Worked example γγ density and resistivity logs

Fig. 6.9 shows a portion of an uncompensated (γγ) density log that is the subject of this worked exercise. There are four depths marked. At each depth, the reader should read the standard density units (SDUs) and note the readings in Table 6.1. A little detective work will be required to determine whether the zones are sandstone or something else. For the density porosity calculation, the log header (see Fig. 6.11) should be inspected to find the correct value of ρ_{mf} to use in the density porosity calculation. (Table 6.1)

Table 6.1 Chapter 6 worked exercise.

Depth	SDU	ρ_B	Porosity for Matrix			R_t	R_{wa}			S_w		
			Sand	Lime	Dol	Ωm	Sand	Lime	Dol	Sand	Lime	Dol
5844												
5885												
5894												
5906												

Fig. 6.10 shows a portion of the resistivity log for the same interval. Note that there are two curves. One is a "guard" or laterolog and the

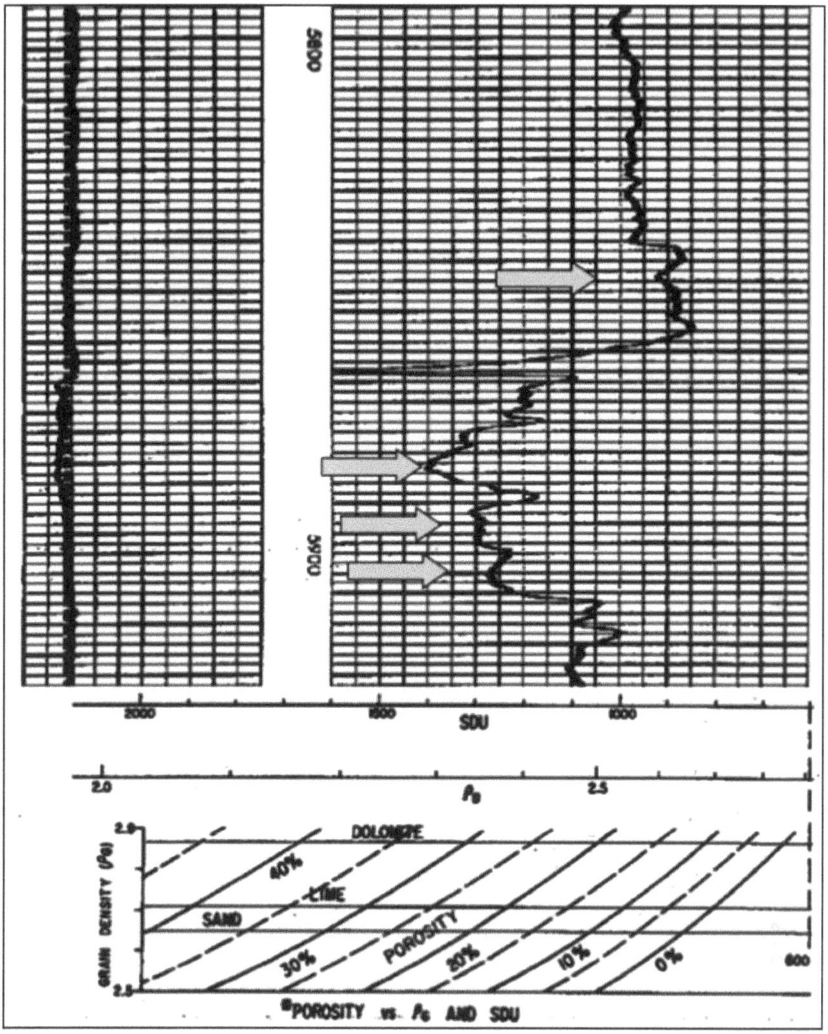

Figure 6.9 Density log for worked example.

other is an induction conductivity curve in millimhos. In the upper, clean (on the gamma ray), interval, the resistivity is quite high. Read the resistivity for all four depths and note them in Table 6.1.

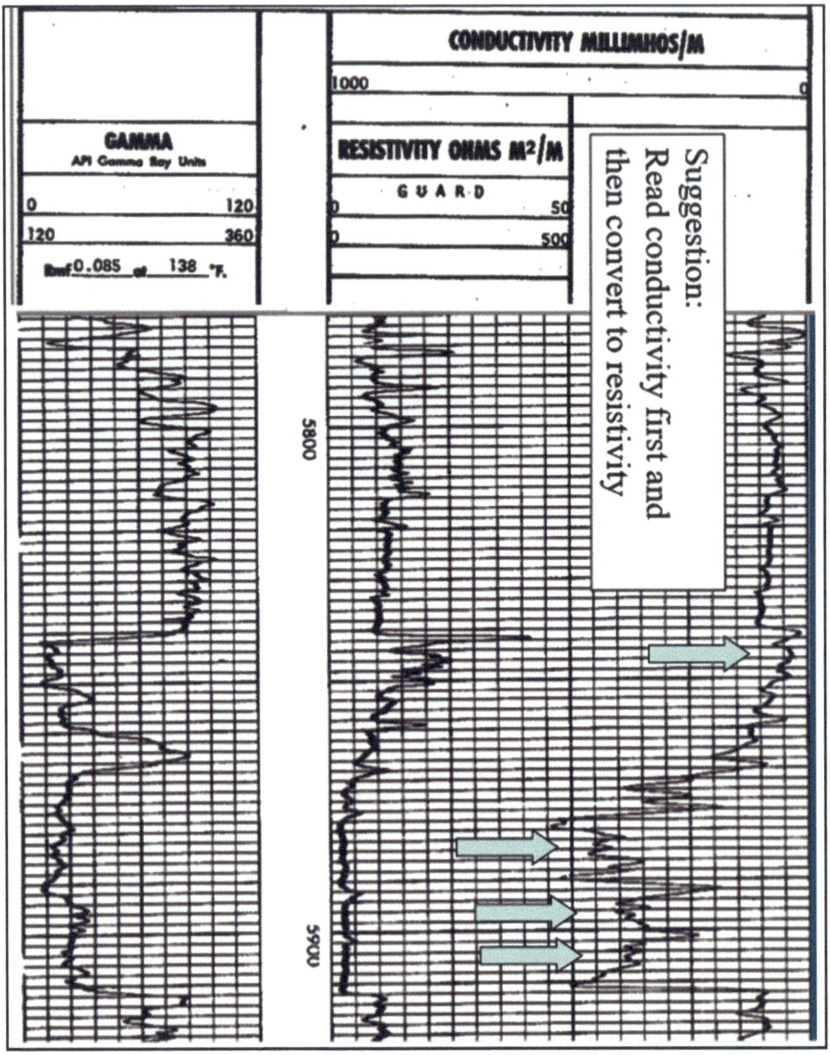

Figure 6.10 Resistivity log for worked example.

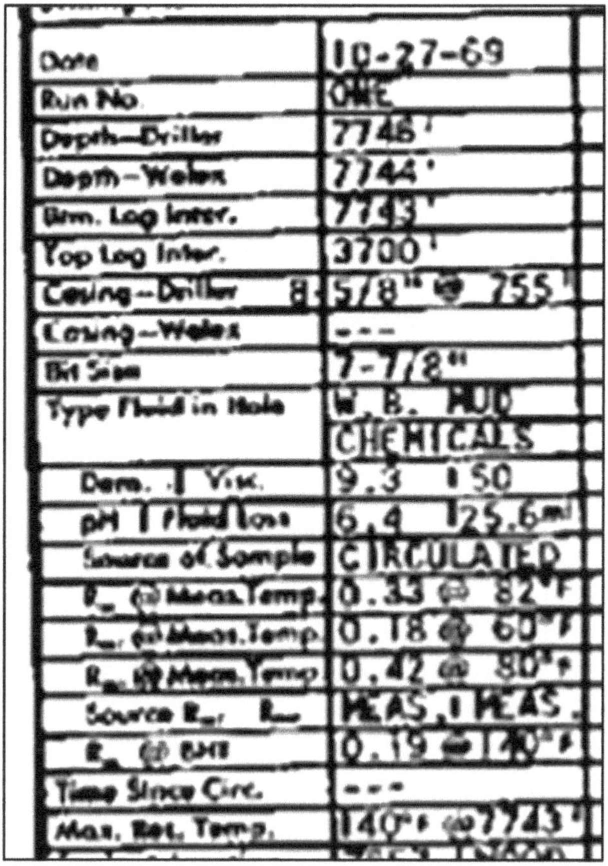

Date	10-27-69
Run No.	ONE
Depth—Driller	7746'
Depth—Welex	7744'
Btm. Log Inter.	7743'
Top Log Inter.	3700'
Casing—Driller	8 5/8" @ 755'
Casing—Welex	---
Bit Size	7-7/8"
Type Fluid in Hole	W. B. MUD CHEMICALS
Dens. | Visc.	9.3 | 50
pH | Fluid Loss	6.4 | 25.6 ml
Source of Sample	CIRCULATED
R_m @ Meas. Temp.	0.33 @ 82°F
R_{mf} @ Meas. Temp.	0.18 @ 60°F
R_{mc} @ Meas. Temp.	0.42 @ 80°F
Source R_{mf} | R_{mc}	MEAS. | MEAS.
R_m @ BHT	0.19 @ 140°F
Time Since Circ.	---
Max. Rec. Temp.	140°F @ 7743'

Figure 6.11 Log header for worked example.

Further reading

Sherman, H., Locke, S., 1975. Effects of porosity on depth of investigation of neutron and density sondes, Paper SPE 5510. Presented at the Annual Meeting, Dallas, September 28–October 1.

Tittman, J., Wahl, J.S., 1965. The physical foundations of formation density logging (Gamma-Gamma). Geophysics 30 (2).

Tixier, M.P., Raymer, L., Hoyle, W.R., 1962. Formation density log applications in liquid-filled holes. J Pet. Tech 15 (3).

Wahl, J.S., Tittman, J., Johnstone, C.W., 1964. The dual spacing formation density log. J. Pet. Tech. 16 (12).

Exercise solutions

a. In Fig. 6.12, the log readings in SDUs and the corresponding values of formation bulk density, ρ_B, are marked.

b. In Fig. 6.13, the formation resistivities of the four zones are marked. It should be noticed that the high-resistivity zone at 5844 has a conductivity of 50 mmhos, which is easier to read off the log and converts to an R_t of 20 Ωm. Note that the gamma ray log has been color coded on the resistivity presentation to highlight the shales (in green) and the clean, porous, zones (yellow).

c. The log header, shown in Fig. 6.11, gives a value of $R_{mf} = 0.18$ @ 60°F. This corresponds to a mildly saline mud system at 46 kppm NaCl. Thus the calculation of density porosity should use the mud filtrate density, $\rho_{mf} = 1.025$.

d. Whatever matrix material is assumed for the lower (deeper) three zones, the density porosity computes with porosities in the 20% – 30% range. The top zone, however, appears to be of lower porosity.

e. The R_{wa} calculations show that the lower three zones show substantially the same low number for R_{wa} in the order of 0.1 Ωm. It can safely be assumed that these three zones are water bearing.

f. However, if the top zone (the shallowest) is considered to be sandstone, the computed water saturation is greater than 1, which is impossible. Indeed even if this top zone is thought to be limestone, the water saturation will still greater than 1. The only logical conclusion is that this is a dolomite, in which case the porosity is 14.2% and the water saturation is 47%.

Table 6.2 Solution to Chapter 6 worked example.

Depth	SDU	ρ_B	Porosity for Matrix			R_t	R_{wa}			S_w %		
			Sand	Lime	Dol	Ωm	Sand	Lime	Dol	Sand	Lime	Dol
5844	900	2.61	0.025	0.060	0.143	20.0	0.013	0.073	0.408	267%	111%	47%
5885	1400	2.30	0.219	0.247	0.313	2.2	0.105	0.134	0.216	**92%**		
5894	1300	2.33	0.200	0.229	0.297	2.5	0.100	0.131	0.220	**94%**		
5906	1275	2.36	0.181	0.211	0.280	2.7	0.089	0.120	0.212	**100%**		

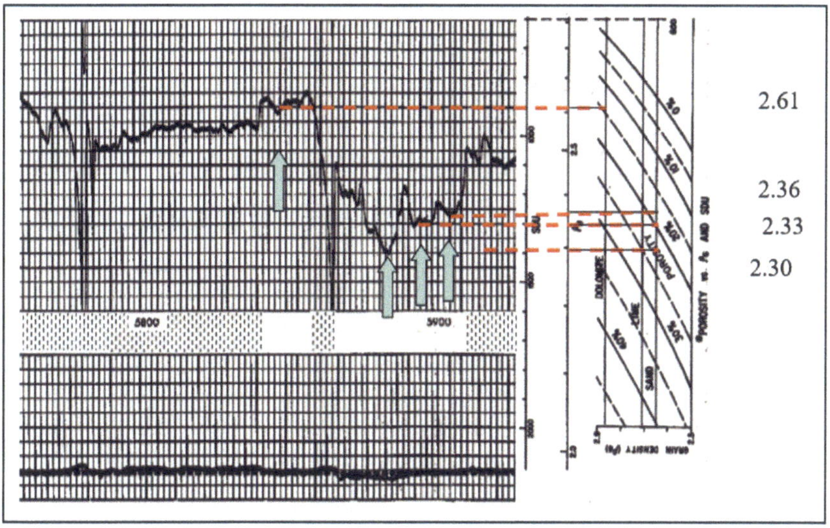

Figure 6.12 Porosity in four zones.

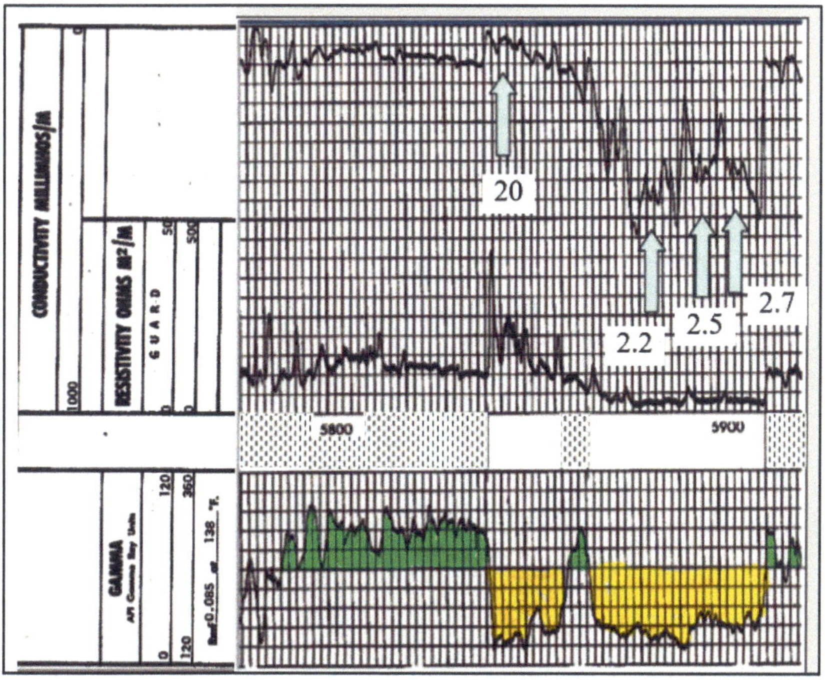

Figure 6.13 Resistivity readings in four zones.

CHAPTER 7

Sidewall Neutron and Compensated Neutron logs

Abstract

Due to the complexity of finding formation porosity from the original "count rate" neutron tools that topic was covered in detail in Chapter 5. Now we turn our attention to all that followed the pioneering neutron logging tools and address the *epithermal neutron* tools and the compensated, (dual-detector) *thermal neutron* tools. In this chapter, we also provide a perspective of all the many combinations and permutations of the types of radioactive source, atomic particles propagated and detected, and the number of detectors and their spacings as used in what we may call the nuclear logging "zoo."

The first important refinery in Corsicana Texas in 1898. *Courtesy Texas Mid-Continent Oil and Gas Association.*

7.1 Epithermal neutron logs

The first pad contact nuclear "porosity" device was the epithermal sidewall neutron tool with the sidewall neutron porosity (SNP) mnemonic. It was introduced in 1962 with a single chemical source of fast neutrons coupled with a single detector of epithermal neutrons. This method of

Formation Evaluation with Pre-Digital Well Logs
DOI: https://doi.org/10.1016/B978-0-12-820232-6.00007-8

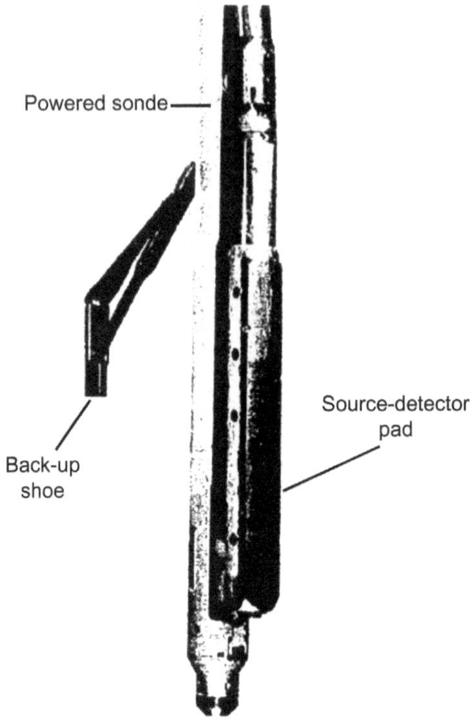

Figure 7.1 Sidewall neutron porosity sonde.

detection avoided the capture gamma rays originating from chlorine atoms in salty formation pore fluids. The epithermal neutron flux thus better represented the hydrogen content of the pore space, which related more directly to the porosity. Fig. 7.1 shows this SNP pad contact tool in a borehole. Note that the powered back-up "shoe" was remotely opened once the tool reached the logging depth and thereafter the tool was raised up the borehole with the combination source-detector pad making hard contact with the formation being logged. Since there was only a single detector in this tool, no allowance was made for any mud cake that might come between the pad and the formation and any corrections for mud cake that had to be made, postlogging, by hand.

One of the redeeming features of the SNP was its ability to measure porosity in both fluid-filled and empty (or air-drilled) holes. Fig. 7.2 shows the log presentation both in mud filled and empty holes. Note that the scaling is primarily in Limestone porosity units and directly in percent. There was no need to make conversions from count rates to porosity. Any postlogging manipulation was restricted to minor mud cake correction and/or for lithologies other than limestone.

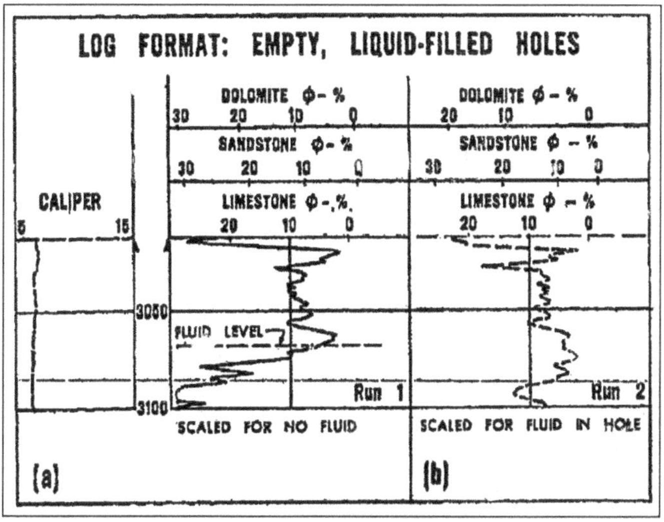

Figure 7.2 Sidewall neutron porosity log scaled for (a) empty hole and (b) fluid-filled hole.

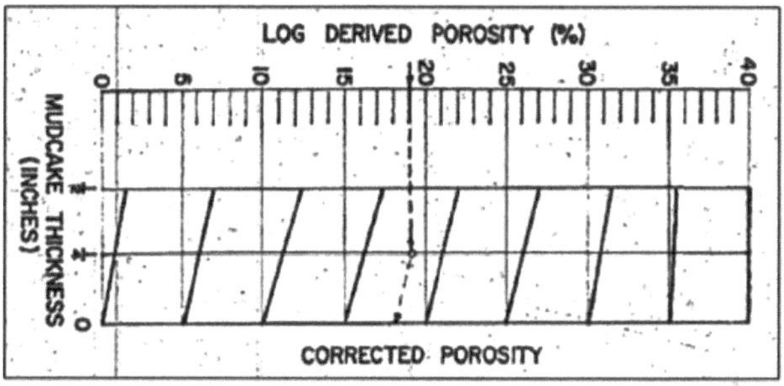

Figure 7.3 Mud cake correction for the sidewall neutron porosity tool.

Corrections for mud cake were applied by the use of a simple chart of the type shown in Fig. 7.3 and amounted to approximately 1 porosity unit (p.u.) per ¼″ of mud cake.

The SNP tool saw its major application in the low porosity carbonates typical of the Permian Basin of West Texas and New Mexico that were being developed in the 1960s. The combination of a linearly scaled density log with a similarly scaled neutron log allowed for a series of crossplots that mapped both porosity and rock type in one place. Fig. 7.4 gives an example of one such plot[1]. This type of plot formed the basis for a whole generation of log analysis methods and routines and laid the groundwork for the burgeoning art of computer processed interpretations that were to follow.

[1] See Chapter 15 for a fuller treatment of lithology identification from old logs.

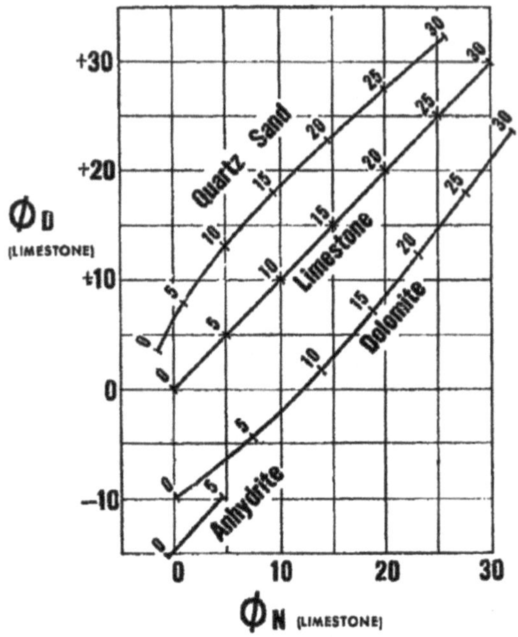

Figure 7.4 Density-Neutron crossplot for porosity and lithology.

Superposition of neutron- and density-derived porosity curves also became a standard for "quick-look" rock typing. If both density and neutron logs were scaled in limestone units, then the two would overlay in limestone but diverge one way in sandstone and the other in dolomite. This is illustrated in Fig. 7.5. The offsets for several matrix materials in listed in Table 7.1.

At the same time, the Dresser Atlas company fielded a combination tool that included a neutron curve (obtained from a conventional epithermal sidewall neutron pad mounted on a mechanical carrier equipped with a hydraulic retractor) and a sidewall acoustic section mounted on the same carrier but contacting the opposite side of the borehole. The entire acoustic section, consisting of a transmitting transducer and two receiving transducers was contained in a pad approximately two feet in length. The receiver spacing was 6 in. producing an acoustic interval transit time curve (Δt) with much sharper vertical resolution than that obtained from a conventional acoustic logging device.

An example of the SNP/sonic combination is given by Fig. 7.6. Note that by judicious scaling of the SNP porosity curve and the Δt sonic travel time curve, the differentiation between the dolomite and anhydrite formations is made evident by visual inspection only.

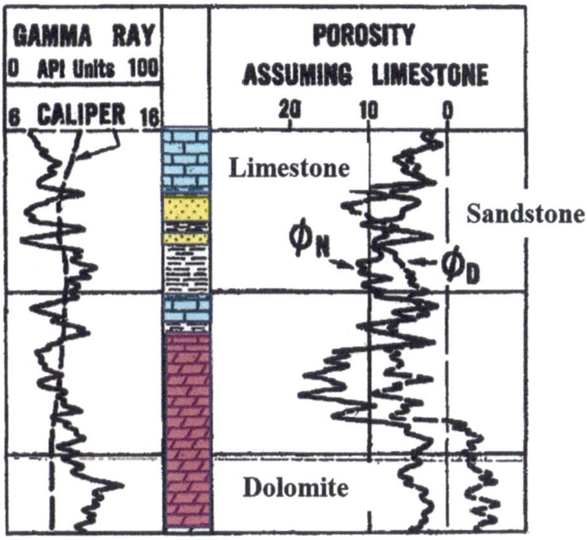

Figure 7.5 Lithology identification from relative ϕ_N and ϕ_D positions.

Table 7.1 Neutron density lithology offsets.

Matrix Lithology	Curve Separation Magnitude p.u.	Curve Separation Direction
Limestone	-	-
Dolomite	~12	$\phi_N > \phi_D$
Sandstone	5 to 7	$\phi_N < \phi_D$
Anhydrite	~15	$\phi_N \approx 0 >> \phi_D$
Shale	5 to 25	$\phi_N > \phi_D$

7.2 Compensated Thermal Neutron logs

The compensated thermal neutron logging tools (widely known as the CNL) were introduced in the late 1960s. As in previous neutron tools, a single chemical (AmBe) neutron source sent out an unfocussed "cloud" of neutrons, but unlike previous neutron tools, the CNL had *two* detectors of thermal neutrons. This design provided a semi-automatic compensation for the effects of mud in the hole and mud cake in front of porous formations. Absolute count rates at the two detectors varied with changing environmental conditions, as well as with the formation porosity; however, the *ratio* of the two count rates proved to be only slightly affected

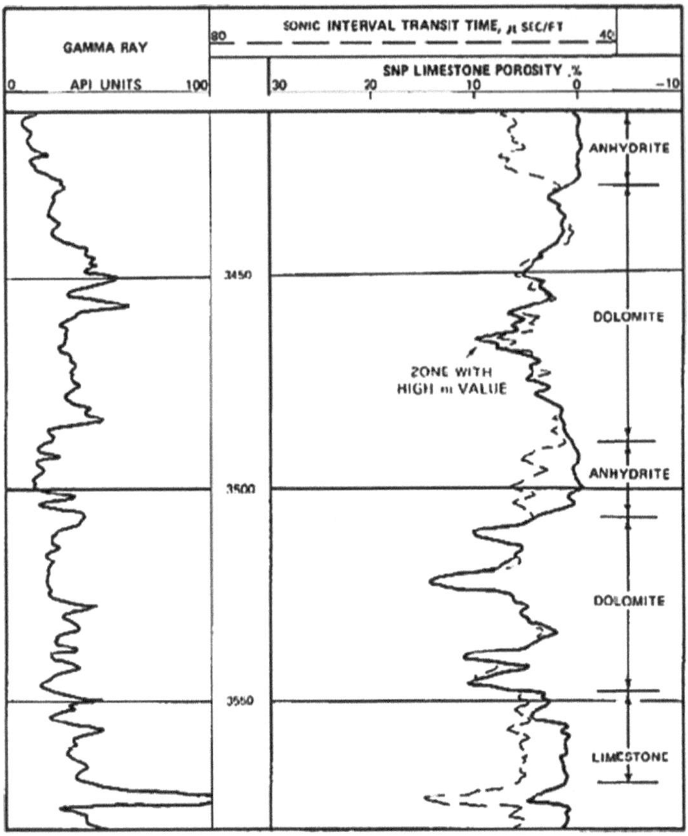

Figure 7.6 Sidewall neutron porosity neutron/sonic presentation in hard rock sequence.

by changes in mud weight, hole size, and so on and was the direct indicator of formation porosity.

Whereas the driver for the epithermal pad contact neutron tools, SNPs, was for better porosity definition in the limestones and dolomites of the Permian Basin, it was the need for a better neutron tool for the higher porosity formations (typical of the Gulf Coast of the United States) that drove the engineering of the dual-spaced thermal neutron tools. To a large extent, these CNLs became the "standard" well into the digital era. Indeed, most of the readers of this work will be familiar with the ubiquitous neutron/density combination, illustrated in Fig. 7.7, that is run today almost as the default "porosity" combo log. The characteristics and usefulness of this type of neutron tool is fully discussed elsewhere in the literature and will not be covered here in any further detail. That it is still in use after 50 years speaks well of its design and usefulness.

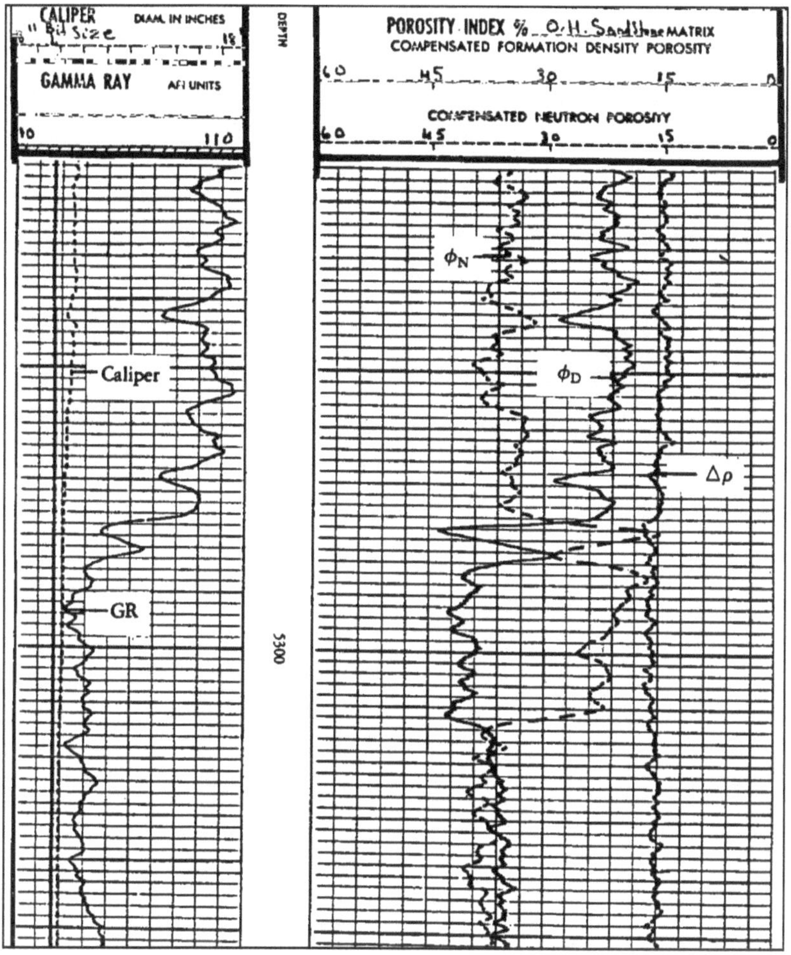

Figure 7.7 "Standard" compensated neutron—compensated density log.

7.3 Neutron logging tool perspective

Fig. 7.8 shows a "generic" neutron logging tool from which the differences between the original "count rate" neutron gamma tools (GNTs), the epithermal neutron tools (sidewall neutron porosities [SNPs]), and the CNLs can be appreciated.

In order to bring some perspective to the nuclear "zoo" the various combinations of radiation sources, particles detected, kinds (and numbers) of detectors have been gathered in Table 7.2.

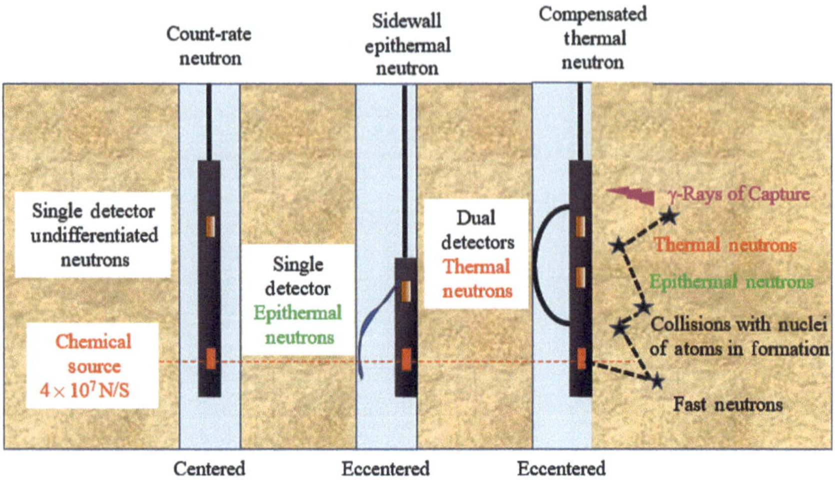

Figure 7.8 Generic neutron logging tool.

Table 7.2 The nuclear "zoo".

Type of Device	Radiation Source	Number of Detectors	Unit Measured	Uses	Remarks
Gamma Ray	Natural	1	γ Rays (API)	Lithology	GR
Neutron-γ	RaBe or PuBe	1	Capture γ's	ϕ, Lith, Gas/Oil	First N Tool
Neutron-Neutron	AmBe	1	Epithermal N's	ϕ, Lith, Gas/Oil	SNP
Gamma Gamma	γ's from Cesium	1	Scattered γ's	ρ B	FDL
Compensated Density	γ's from Cesium	2	Scattered γ's	ρ B	FDC
Neutron-Neutron	AmBe	2	Thermal N's	ϕ, Lith, Gas/Oil	CNL

Further reading

Belnap, W.B., Dewan, J.T., Kirkpatrick, C.V., Mott, W.E., Pearson, A.J., Rabson, W.R., 1960. API Calibration for nuclear logs. Drill and Prod Practice. API, New York.

Gilchrist, W.A., 2009. Compensated Neutron Log Response Issues. Society of Petrophysicists and Well-Log Analysts.

Grosmangin, M., Walker, E.B., 1957. Gas detection by dual-spacing neutron logs in the greater oficina area, Venezuela. J. Pet. Tech. 9 (5), 140–147.

Lebreton, F., Youmans, A.H., Oshry, H.L., Wilson, B.F., 1963. Formation evaluation with nuclear and acoustic logs. Trans SPWLA Fourth Annual Logging Symposium, Paper IX.

Stroud, S.G., Schaller, H.E., 1960. A new nuclear log for the determination of reservoir salinity. J. Pet. Tech. 12 (2), 37–41.

Tittman, J., Sherman, H., Nagel, W.A., Alger, R.P., 1965. The sidewall epithermal neutron porosity log. Paper SPE 1180, Presented at the 40th SPE Annual Meeting. 3–6 October.

Tittman, J., Sherman, H., Nagel, W.A., Alger, R.P., 1966. The sidewall epithermal neutron porosity log. J. Pet. Tech. 18 (10), 1180.

Traugott, M.O., 1970. Log Evaluation of a Heterogeneous Carbonate Reservoir Cato San Andres Field. Society of Petrophysicists and Well-Log Analysts, SPWLA 11th Annual Logging Symposium, May 3-6.

CHAPTER 8

Microlog recordings and interpretations

Abstract

For the modern analyst, faced with the task of performing formation evaluation with pre-digital well logs, the microlog is an unfailing crutch. Its value lies in its simplicity of electro-mechanical design and operation and the ease of interpreting its recordings. At the purely visual level the microlog points to the porous and permeable zones in a logged section of borehole. Basic petrophysical (read "Archie") principles then lead directly to fairly reliable estimates of porosity. By extending the analysis chain a little further the microlog provides us with the value of R_{xo} without running a conventional resistivity log at all.

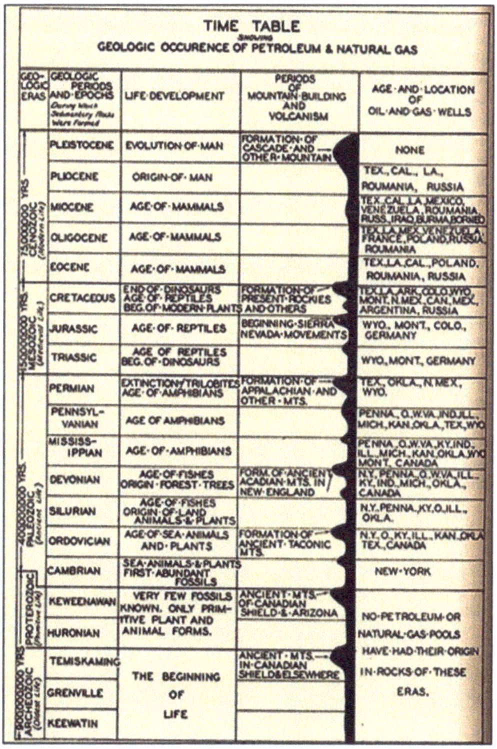

An early petroleum geologic column

Anon

DOI: https://doi.org/10.1016/B978-0-12-820232-6.00008-X

8.1 The tool and its measurements

The microlog is a pad contact tool that came on the open-hole logging scene in 1949 and became a staple of logging suites for that era. It is a simple, yet effective, device that offers two resistivity recordings from which the log analyst may deduce both porosity and permeability. The tool achieves this by the simple expedient of using a pad with three electrodes embedded in it to record the resistivity of the flushed zone, R_{xo}, and that of the invaded zone, R_i. Fig. 8.1 shows a schematic of the tool and Fig. 8.2 a typical microlog recording.

The electrode arrangement is a standard electric log with A and M electrodes. In some logs the recorded curves will be labeled as "micro-normal" and "micro-inverse," in others as $R_{1'' \times 1''}$ and $R_{2''}$, and in yet others as "normal" and "lateral."

In the example shown in Fig. 8.2, the places where the two resistivity curves are substantially similar, but separated by a small resistivity difference, have been highlighted in red. This indicates that the formation in front of the pad at those depths was both porous and permeable.

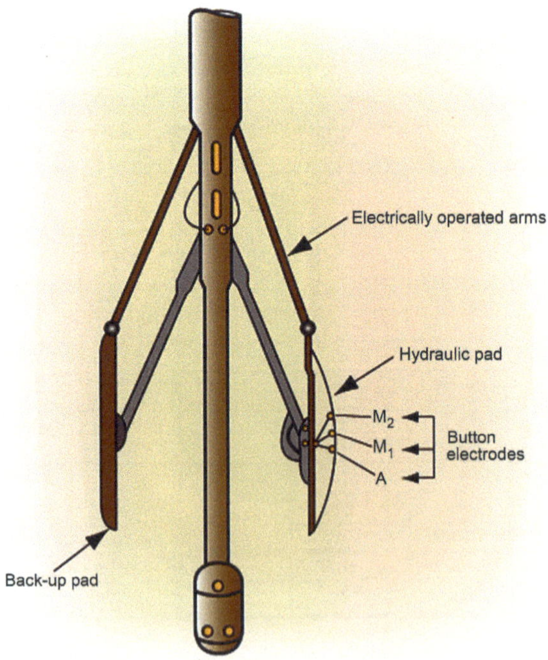

Figure 8.1 The microlog tool. *Image courtesy Schlumberger.*

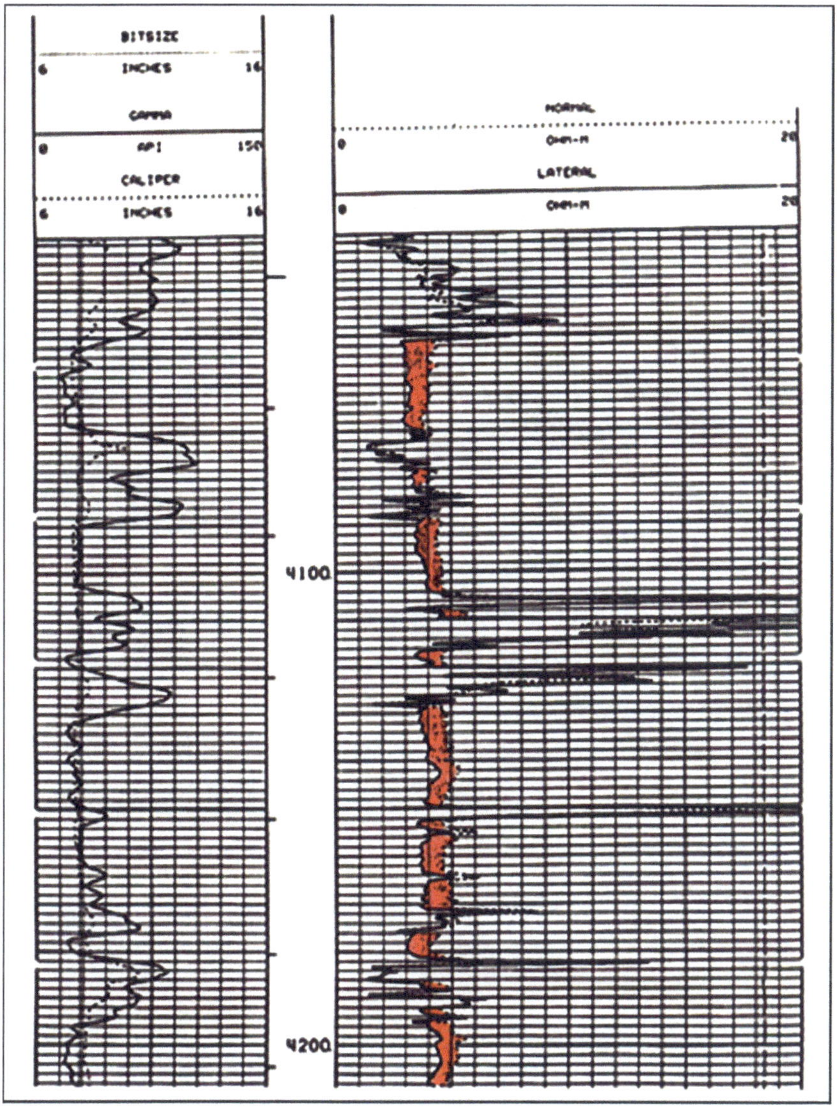

Figure 8.2 A typical microlog recording. *Image courtesy Schlumberger.*

When drilled with over balanced water-based mud systems porous and permeable formations are subjected to a mini-water flood. Mud cake builds up on the borehole wall and mud filtrate replaces movable fluids in the near-wellbore annular volume. Thus water saturation changes radially away from borehole. Consequently formation resistivity also changes radially.

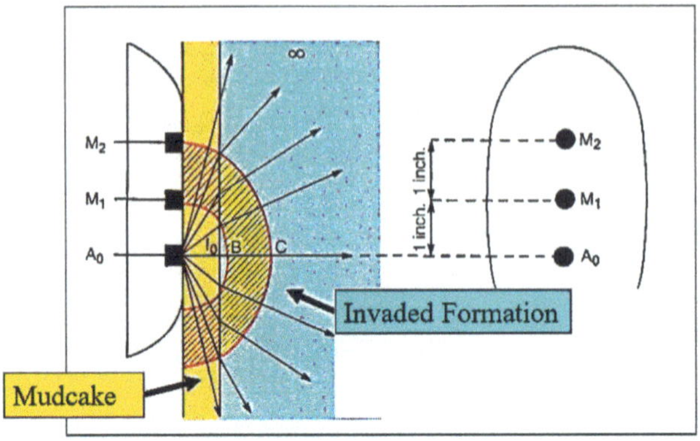

Figure 8.3 Closeup of the microlog current flow in the near-wellbore invaded formation. *Image courtesy Schlumberger.*

In Fig. 8.3 it can be seen that current from electrode A_0 to M_1 passes mostly through the mud cake (color orange) while current from A_0 to M_2 passes though part of the invaded zone (color yellow) as well.

8.2 Causes and effects of invasion

The invasion of mud filtrate is controlled by a number of factors including water loss, time, formation porosity, and permeability. Log analysts use special nomenclature to label the near-wellbore geometry, resistivities, and saturations. It is assumed that the reader is familiar with this nomencalture which is availble in standard texts such as Bateman (2012). Fig. 8.4 is offered as a reminder.

8.3 Porosity from microlog

An empirical chart may be used to deduce porosity from a reading of the resistivities of the two microlog curves and knowledge of the mud resistivity. The log heading will list the value of the mud resistivity, R_m, at its measure temperature (normally the ambient surface temperature in the logging truck) and this value needs to be converted to a resistivity at the higher temperature found at logging depth. For this purpose the well-known (Arps) formula may be used as follows:

$$R_2 = R_1 \frac{(T_1 + 7)}{(T_2 + 7)}$$

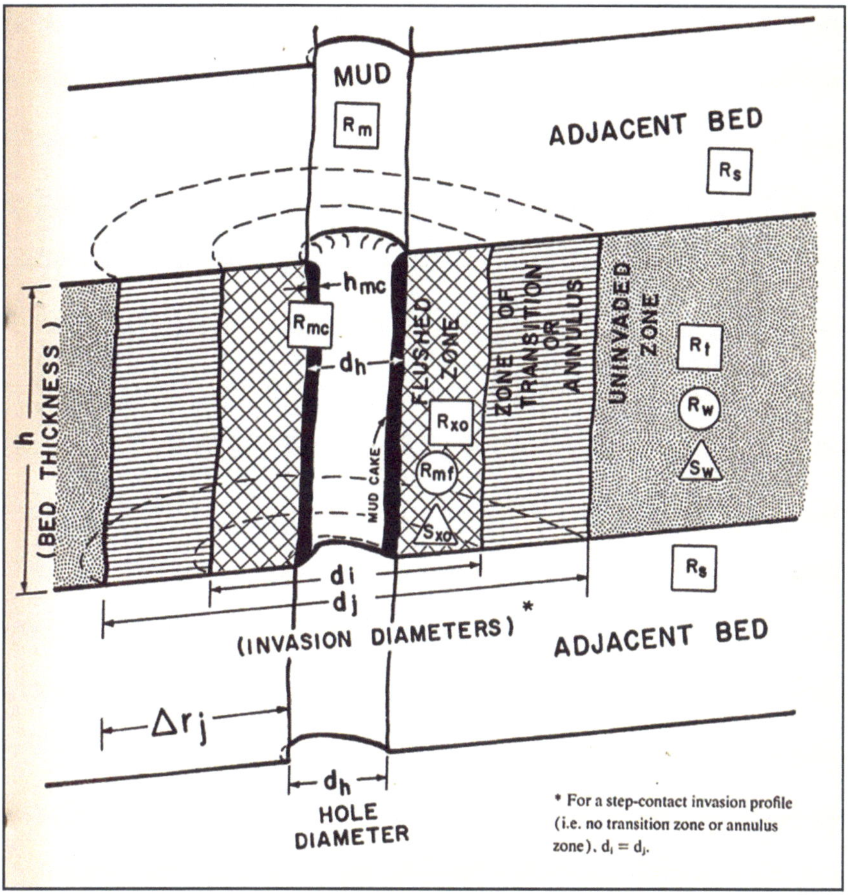

Figuure 8.4 Invaded formation schematic. *Image courtesy Schlumberger.*

where temperatures are in degrees F and resistivities are in Ωm. By way of illustration, the chart shown in Fig. 8.5a may be used. R_m, at formation temperature, is entered on the *Y*-axis and $R_{2''}$ from the log, is entered on the *X*-axis. At the intersection of the chosen data points the value of *F*, the formation factor, may be read. From there a second step can transform the *F* value to a porosity value using a relationship of one's choice in the form of

$$F = \frac{a}{\phi^{m}}$$

The conversion of *F* to ϕ can be accomplished either by use of a calculator or, as was common practice in the pre-digital era, when calculators were not available, by use of the sub-chart shown as Fig. 8.5b.

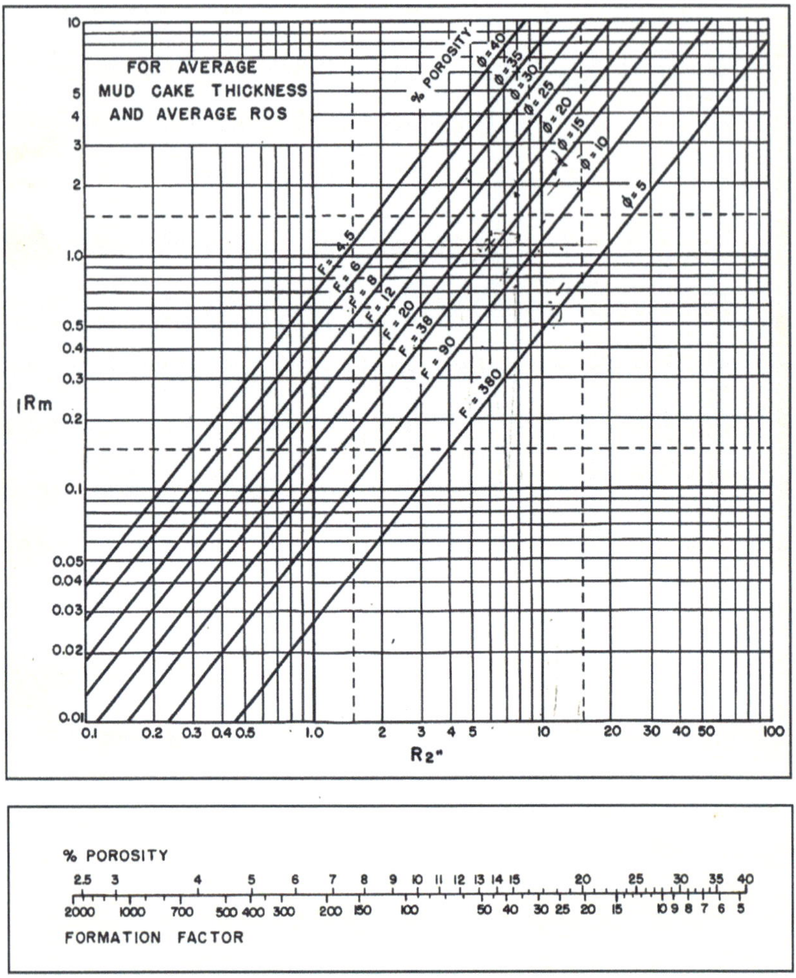

Figure 8.5 a—Empirical microlog chart for porosity, b—Conversion of formation factor F, to porosity ϕ. *Image courtesy Schlumberger.*

Exercise 8.1

Given that $R_m = 0.6\ \Omega m$ and $R_{2''} = 3\ \Omega m$

Read F from Chart 5.5a and convert to ϕ

Solution at the end of Chapter

8.4 R_{xo} from the microlog

An alternative interpretation chart, again, empirical, allows the analyst to derive a value for the invaded formation resistivity, R_{xo}, from which the value of the water saturation in the invaded zone, S_{xo}, may subsequently be derived by application of Archie's equation.

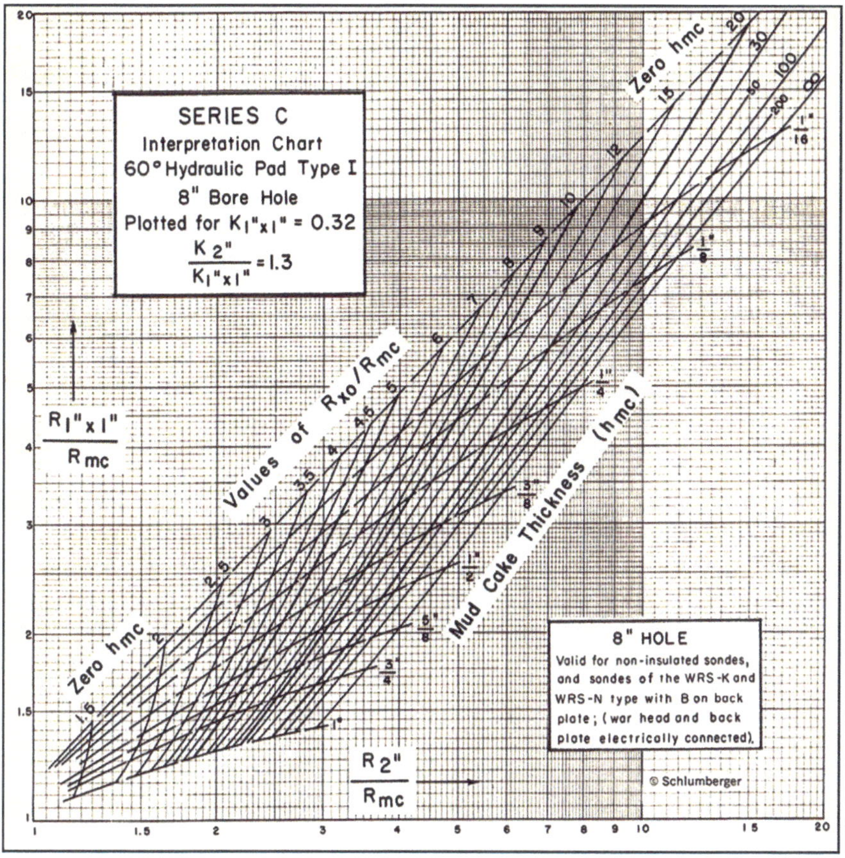

Figure 8.6 Alternative empirical microlog interpretation chart. *Image courtesy Schlumberger.*

Referring to Fig. 8.6, it may be observed that the *Y*-axis plots the values of the fraction $R_{1'' \times 1''}/R_{mc}$ and the *X*-axis plots the values of the fraction $R_{2''}/R_{mc}$. The intersection of lines extended from the plotted points on the *X*- and *Y*-axes defines a point on the chart where the value of the fraction R_{xo}/R_{mc} may be read along with the corresponding mud cake thickness, h_{mc}.

Exercise 8.2

Given that:

$R_{mc} = 1.0\ \Omega m$

$R_{1'' \times 1''} = 2.3\ \Omega m$

$R_{2''} = 3\ \Omega m$

Find mud cake thickness, h_{mc} and the value of the invaded zone resistivity R_{xo}

Solution at the end of Chapter

Once the value of R_{xo} has been derived from Fig. 8.6 it may be combined with the value of the mud filtrate resistivity, R_{mf}, to calculate S_{xo} following the Archie equation[1]:

$$S_{xo} = \sqrt[n]{\frac{F\,R_{mf}}{R_{xo}}}$$

Exercise 8.3

Given that $\phi = 0.25$

Assume that R_{mf} at formation temperature is 0.35 Ωm

Use the value of R_{xo} found in Exercise 5.2

Calculate S_{xo} using the Archie constants $a = 1$ and $m = n = 2$.

Solution at the end of Chapter

In pre-digital formation evaluation the value of S_{xo} carried importance for several reasons. A good value of the water saturation in the uninvaded formation, S_w, was not always readily available, either for lack of a reliable measure of porosity, or for lack of an accurate value for R_t. Given these uncertainties it was common to rely on an empirical observation that S_w and S_{xo} may be related by a power law equation of the sort

$$S_{xo} = S_w{}^{0.2}$$

Such a relationship can be used in the reverse to determine a value of S_w given the S_{xo} value derived from the microlog. In that case the equation would read[2]:

$$S_w = S_{xo}{}^{5}$$

Exercise 8.4

Convert the S_{xo} derived in Exercise 8.3 to a value for S_w

Solution at the end of Chapter

Further reading

Doll, H.G., 1950. The microlog - a new electrical logging method for detailed determination of permeable beds. Trans. AIME 189, 155.

Hilchie, D.W., 1979. The microlog. Old Electrical Log Interpretation, 61–70.

[1] It should be noted that the value of R_{mf} can be found by reading the log heading where R_{mf} at surface temperature is duly noted. This surface measured value can then be transformed to its equivalent at formation temperature, as before, via the Arps formula.

[2] In both formulae it goes without saying that the saturations are expressed as factions of 1.

Tixier, M.P., 1956. Fundamentals of electrical logging - microlog and microlaterolog. In: Fundamentals of Logging. Univ. Kansas, Petroleum Eng. Conf. 2 and 3.

Exercise solutions

Exercise 8.1

$F = 20$

$\phi = 20.2\%$

Exercise 8.2

$h_{mc} = 3/8''$

$R_{xo} = 7\ \Omega m$

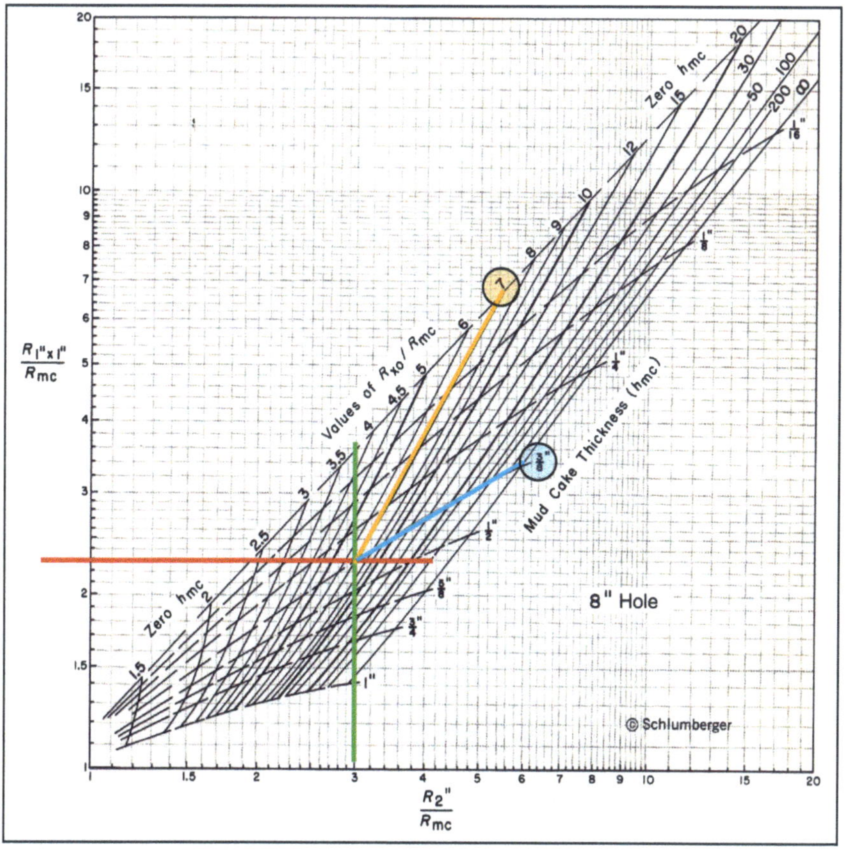

Exercise 8.3

$\phi = 0.25$

$F = 16$

$R_{xo} = 7\ \Omega m$

$S_{xo} = 89\%$

Exercise 8.4

$S_w = S_{xo}^5 = 0.89^5 = 0.56$ (or 56%)

CHAPTER 9

Spontaneous potential

Abstract

The spontaneous potential (SP) is a player who takes on many roles in the drama of well log interpretation. Originally, as the joker, it was just a nuisance to the main formation resistivity play—but not for long—for soon its true worth was realized as a predicter of porous and permeable zones. When yet another veil was removed it was the messenger for both R_w and V-shale but not before the geologists were to sing its praises for revealing sedimentary patterns. It was even co-opted into a brief romance with dip metering. SP, surely, stands for *special*.

Parasols to protect men and apparatus in the California desert, 1932
Image courtesy Schlumberger.

9.1 Originally called "porosity log"

Five years after the first electric log in Alsace, a spontaneous potential (SP) and resistivity log was recorded by Mr. William Gillingham in Texas on which is clearly labeled the SP curve as "porosity" as shown in Fig. 9.1. Neat call-outs in the depth track denote the depths where there is sand,

Formation Evaluation with Pre-Digital Well Logs
DOI: https://doi.org/10.1016/B978-0-12-820232-6.00009-1

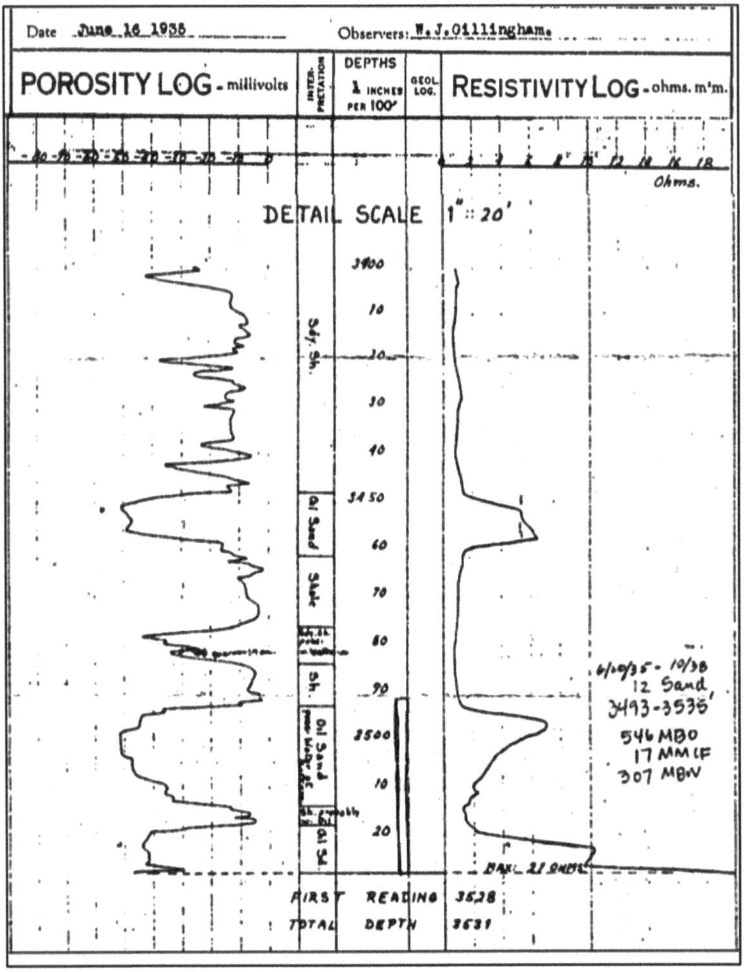

Figure 9.1 Early SP labeled as "porosity log."

shale, oil sand, and so on. The scaling of the "porosity" track is in millivolts which is, indeed, the units in which the curve was recorded.

9.2 The SP combined with core analysis predicts porosity

Certainly, in the early days of electric logging, the SP was looked on as a porosity indicator. One of the techniques used was to key the SP deflection to porosity in some "anchor" formation in which coring had afforded a direct measurement of porosity. Porosity in adjacent sands was then calculated using a ratio between the maximum SP in thick clean sand (SSP)

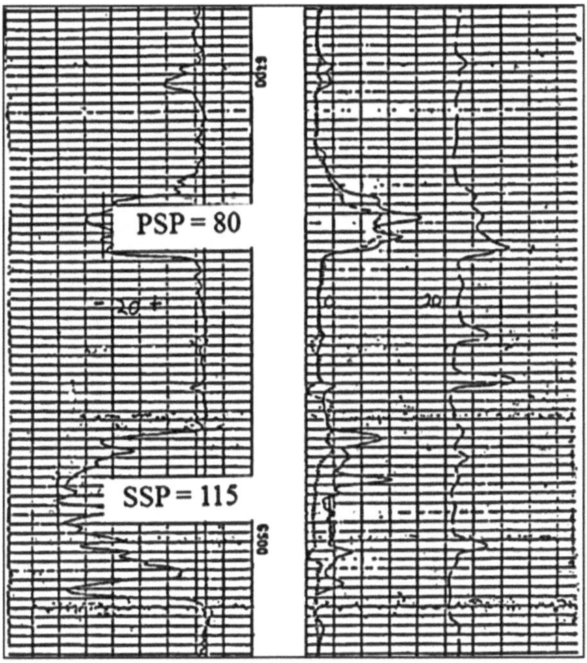

Figure 9.2 Porosity from SP.

and the SP in the bed of interest (PSP). The ratio was given the symbol α, such that

$$\alpha = \mathrm{PSP}/\mathrm{SSP}$$

Then if ϕ_{Max} is the clean sand porosity (from core) then the porosity in the neighboring sands could be calculated from the formula

$$\phi = \frac{\phi_{\mathrm{Max}}}{(2 - \alpha)}$$

By way of example the method may be illustrated from the log shown in Fig. 9.2.

In the lower sand the SSP is 115 mV. The porosity there is known to be 32.5% from core analysis. The PSP, read in the upper sand, is 80 mV. Hence the value of α is $80/115 = 0.7$. Applying the algorithm ϕ_{Max} / $(2 - \alpha)$ gives $32.5/(2 - 0.7) = 25\%$.

9.3 SP as an aid to determining R_w

It will be assumed that the reader is familiar with the standard procedures for finding R_w given measurements of R_{mf}, formation temperature and

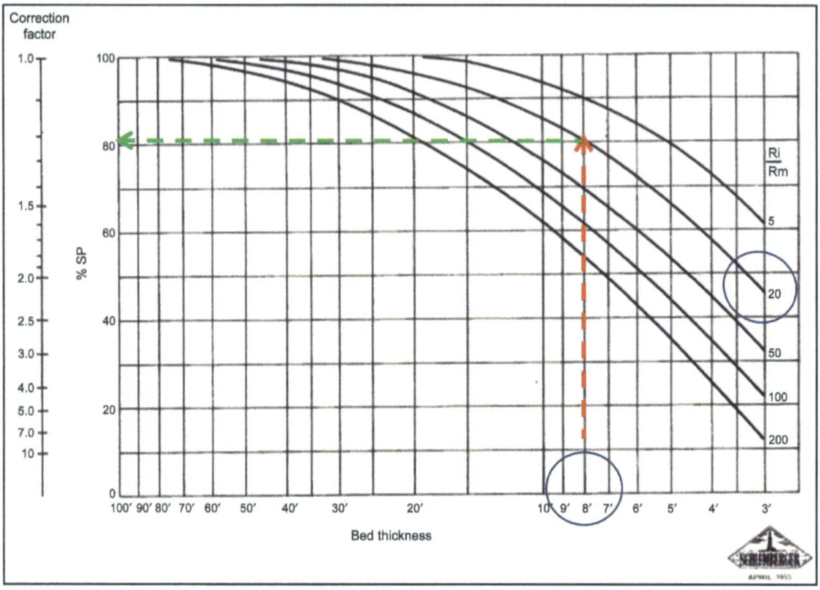

Figure 9.3 Empirical SP correction chart. *Image Courtesy Schlumberger.*

the SP mV corrected for bed thickness[1]. What is worth underlining here are some of the original charts used for this most important task. Fig. 9.3 shows the empirical SP correction chart from 1955 that requires the input of the bed thickness (on the *X*-axis) and a knowledge of the ratio of R_i to R_m, which can be sourced from the resistivity log(s).

The example shown on Fig. 9.3 is for an 8′ thick bed (red line) where R_i is 10 Ωm and R_m is 0.5 Ωm giving an R_i/R_m ratio of 20. The indicated SP correction factor (green line) is 1.25, that is, the conditions of bed thickness and formation and mud resistivities have combined to suppress the full development of the SP to only 80% of the value it would have had if the bed had been infinitely thick. There are, of course, other methods of determining R_w and they should be considered as well. Chapters 11 and 13 should be consulted for confirmatory, independent, R_w estimates.

Fig. 9.4 reproduces a long-forgotten chart published by Elgen in 1956 for determining R_w given three inputs SP, R_{mf}, and K which are related to formation temperature by $K = (T°F + 505)/8$.

[1] A handy reference for this is Chapter 9 of "openhole log analysis and formation evaluation." Additionally, a logic flow chart for R_w computation is to be found in the Appendix to this chapter.

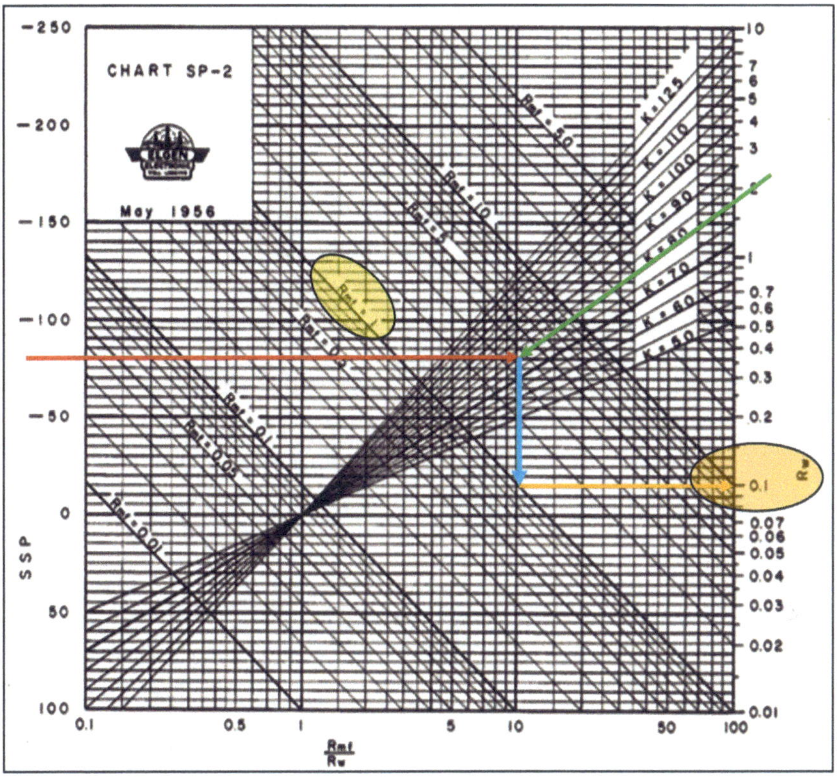

Figure 9.4 R_w from SP using Elgen chart.

As an example, we will take the case where

$$\begin{aligned} SP &= -80 \text{ mV} \\ T &= 135\ ^{\circ}\text{F} \\ R_{mf} &= 1\ \ \Omega\text{m} \end{aligned}$$

To use the chart a first step is to enter with the SP in mV from the Y-axis (the red line) to meet with a value of K (the green line). In the case illustrated the SP is −80 mV and K is 80 (135°F). From the point on the chart where the SP and K lines meet a line is drawn down the page (the blue line) to intersect the line of $R_{mf} = 1.0\ \Omega$m. From the intersection of the blue line with the R_{mf} diagonal a horizontal line (the orange line) is drawn to the right of the chart to intersect the R_w axis at the R_w value of 0.1 Ωm. A refreshing, simple, and direct method of graphically solving an otherwise messy algebraic manipulation!

A more exhaustive method for calculating R_w from the SP is given in the Appendix to this chapter in the form of a logic flowchart, originally published in the late 1970s, as the introduction of hand-held programmable calculators allowed the log analysts a degree of freedom from nomograms and hand plotting on paper charts.

9.4 SP as a shale indicator

Log analysts like to use several raw log measurements to quantify the "shaliness" of a formation. Chief among these are the SP and the gamma ray. Given that the electrochemical drivers, that give rise to an SP in the first place, remain unchanged over a logged interval then the SP is a fair indicator of the shale content of a quartz sandstone formation. However, where the driller changed out the mud system between one log run and the next then the SP deflection will mirror not only changes in shale content but also changes in the salinity of the mud filtrate with a consequent change in the SSP. Discounting such changes the standard expression applies to calculate V_{Shale} from the SP as:

$$V_{\text{Shale}} = 1 - \frac{\text{PSP}}{\text{SSP}}$$

On the log shown in Fig. 9.5 two dashed red lines delineate the "clean sand" and "shale." Sand A is considered "clean" and the SSP is 100 mv. Sand B is slightly shaly and has a reduced PSP of only 76 mV. By application of the formula V_{Shale} is given by $1-76/100 = 0.24$ or 24%.

Other indictors of shaliness are also recommended because other factors may reduce the SSP for reasons that have nothing to do with shaliness, hydrocarbon saturation for example (see later). For this reason, it is wise to include a gamma ray shaliness indications (where available) in the analyst's calculations, although even the gamma ray can be misleading in sands that contain potassium feldspar (and therefore appear shalier than they really are to the GR). To confuse matters further some clays are not radioactive (such as kaolinite) and so do not figure as "shaly" to the GR and lead to underestimation of V_{Shale} from the GR.

9.5 SP as a hard rock/ soft rock indicator

In hard rock country the SP is not as predictable as it is in sand/shale sequences elsewhere. The small currents that are generated by membrane potentials and the liquid-junction effects are restrained to flow in the borehole itself due to the high formation resistivity in the low porosity carbonate formations traversed by the bit. In turn, this generates "straight line"

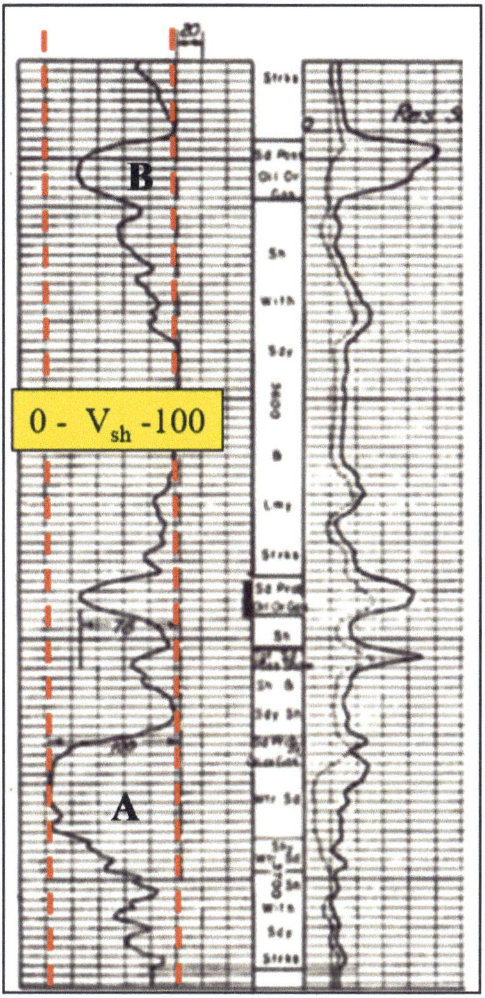

Figure 9.5 SP as shale volume indicator.

portions on the SP traces, only to change direction when coming to either a permeable sandstone or a shale layer. Fig. 9.6 illustrates the phenomenon.

9.6 SP reduction in hydrocarbon-bearing zones

Where the mud filtrate and the connate water interact in the region of the formation immediately adjacent to the borehole the liquid-junction potential is created, and its magnitude will depend on the contrast between the salinity of the two solutions. If some of the connate water has been displaced by hydrocarbons, then the potential generated will be

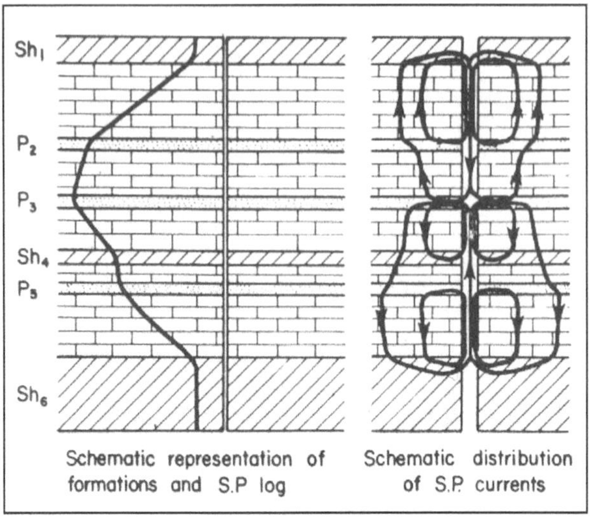

Figure 9.6 SP behavior in low porosity (tight) sequences. *Image Courtesy Schlumberger.*

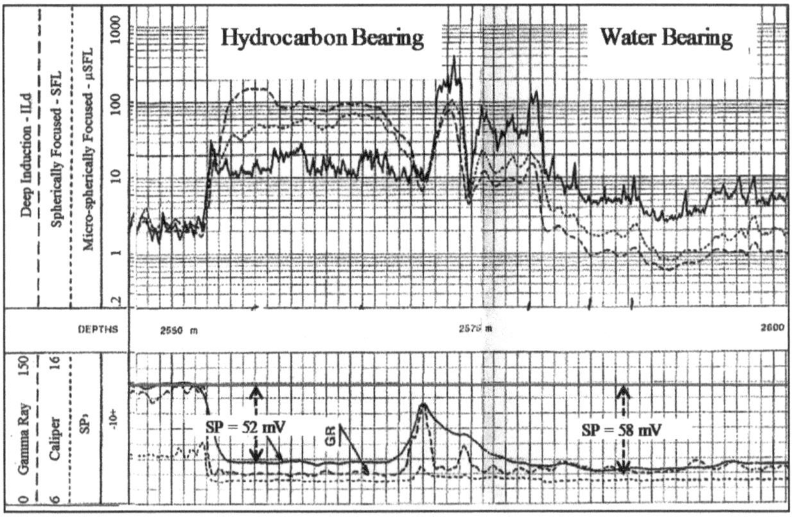

Figure 9.7 SP reduction in hydrocarbon-bearing zone.

diminished. The logs shown in Fig. 9.7 demonstrate this effect. In the lower portion of the logged section (below 2575 m) the formation is clean and wet and the SP deflection from the shale base line (red line) is 58 mV. In the shallower section, immediately below 2553 m, the formation is hydrocarbon bearing and the SP is observed to deflect only 52 mV from the base line. The "hydrocarbon effect," in this case, amounts to a 10% reduction in the SP.

9.7 SP use for quick-look water saturation

In Chapter 13 several "quick-look" methods are documented that allow the analyst simple and quick methods of finding "answers" without excessive log reading and plotting. Amongst those is the SP with a scaled "R_{xo}/R_t" ratio that was popular in the pre-digital era. Fig. 9.8 gives an example of such a plot.

The logic behind this presentation is the same as is used in the "Rocky Mountain method" that is fully explained in Chapter 13. Here the reduction in water saturation as the logging tool passes from the shale and then water bearing sections (below 3940′ depth) is made visually

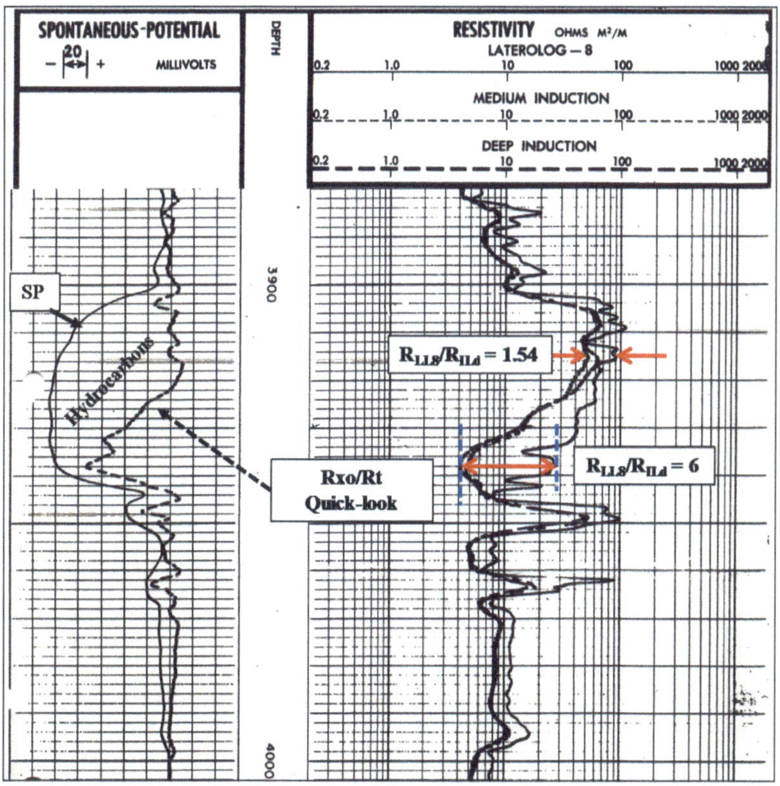

Figure 9.8 $R_{xo}/R_t/R_t$ versus SP curve overlay for quick-look hydrocarbon detection.

evident by the separation between the SP curve from the dashed R_{xo}/R_t curve. The underlying driver of this simple, yet highly effective, display is the actual resistivity reading in the wet and hydrocarbon-bearing intervals. Note that, for example, the ratio R_{LL8}/R_{Ild} at 3938′ is 6.0. However, at 3914′ the same ratio is reduced to 1.54. Many such displays will be found in the archives. Some may even have been enhanced by coloring of the area where the SP and scaled R_{xo}/R_t curves separate.

9.8 SP as a depositional environment indicator

Finally, we should not leave our friend the SP without mentioning the characteristic shapes of SP recordings as they pass through sediments of differing origins. Fig. 9.9 shows the "bell" (fining upward), "the cylinder,"

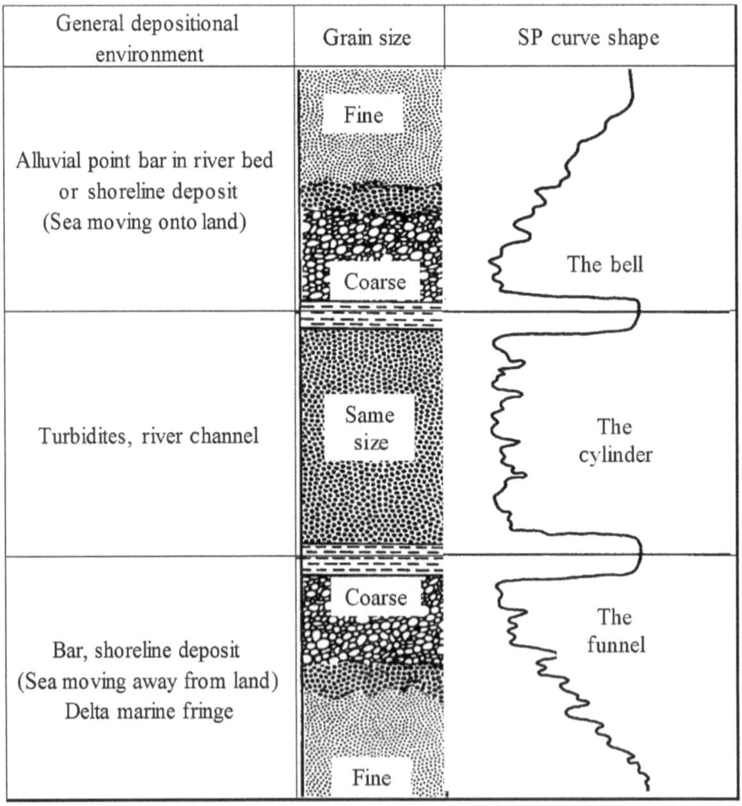

Figure 9.9 SP curve shapes as indicators of sedimentary features.

and "the funnel" (coarsening upward) help in determining the sedimentary environment.

Of all the petrophysical parameters that were measurable (and recorded) in the pre-digital era the SP stands out as the most ubiquitous. The analyst charged with making sense of the old E-logs will find that the SP was recorded in just about every well drilled with fresh mud. It is a good starting point for focusing attention on potential zones of interest. Following the pointers, given in this chapter, many other key "answers" may be derived from this simple, naturally occurring tiny voltage arising from currents flowing in the mud column which, in turn, are due to such key items as porosity, permeability, rock type, water salinities, and even water saturation. Although, today, the SP may be depreciated, in fact it is something special.

Further reading

Bateman, R.M., Konen, C.E., 1977. The log analyst and the programmable pocket calculator- part 1 Rw from the SP. Log Analyst. 18 (5).

Dickey, P.A., 1944. Natural potentials in sedimentary rocks. Trans AIME 155, 30.

Doll, H.G., 1948. The SP log: theoretical analysis and principles of interpretation. Trans AIME 179, 146.

Mounce, W.D., Rust Jr., W.M., 1944. Natural potentials in well logging. Trans AIME 155, 47.

Wyllie, M.R.J., 1949. A quantitative analysis of the electro-chemical component of the SP curve. Trans AIME 186, 17.

Appendix—Flowchart for R_w computation

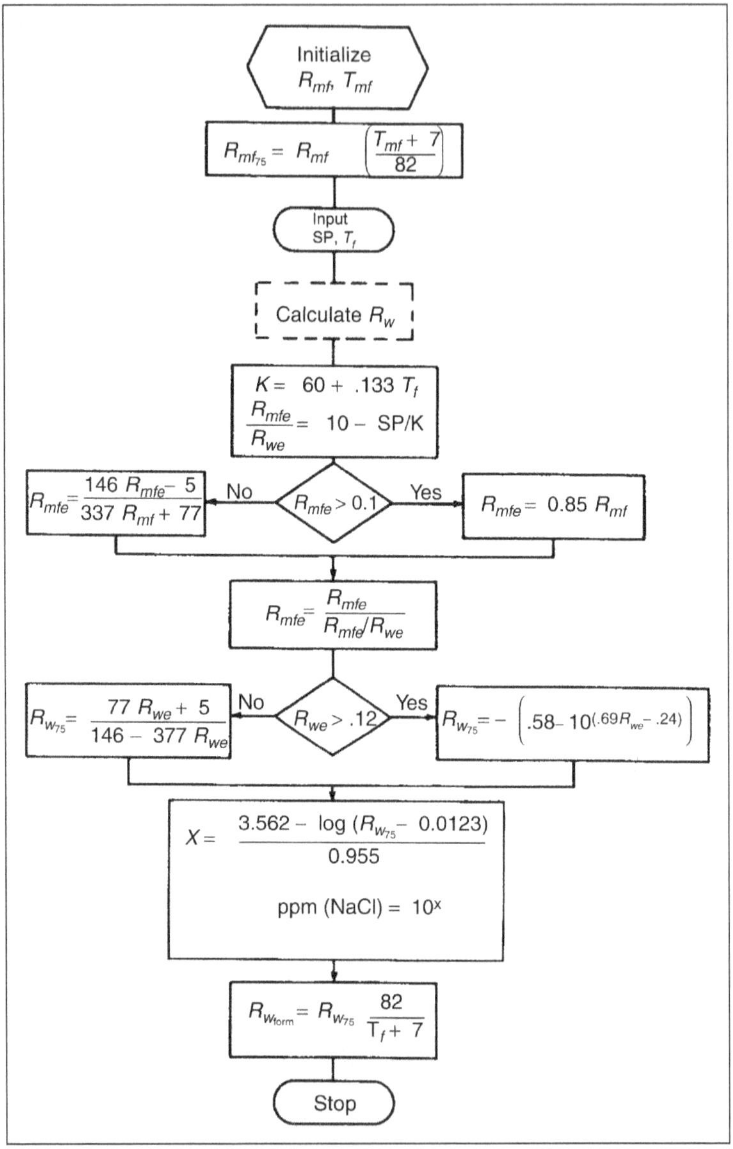

SECTION IV

Saturation Determination and Archie Alternatives

CHAPTER 10

Rocky Mountain Method

Abstract

In this chapter, we get acquainted with a powerful log interpretation technique that not only helped early log analysts find water saturation from a very limited set of log measurements but also set an example for later developments both in logging tool design and in formation evaluation. The pioneer, Maurice Tixier, cleverly saw that if a calculation required an unknown parameter the need for it could be sidestepped by the simple expedient of taking ratios of known or measured parameters. The principle can even be applied to find porosity from a compensated thermal neutron log by taking the ratio of the near and far count rates. The absolute counts depend on both porosity and the unknown effects of hole size and mud weight, but the ratio of the two is practically unaffected by anything but porosity. Tixier's seminal paper was published in 1949 and has certainly stood the test of time.

Water fall by Escher.

DOI: https://doi.org/10.1016/B978-0-12-820232-6.00010-8

10.1 Basic Rocky Mountain method

Maurice Tixier came up with an ingenious method of finding the water saturation from three simple and readily available inputs: R_t, R_{xo}, and SP. The method sidestepped the need to know the porosity or, indeed, to know the value of R_w. In the early days of formation evaluation this was a great plus since there were so many unknowns still to be learned in any given field. Tixier's method relied on assumptions about the relationship between S_w and S_{xo} in water-based mud filtrate invaded formations. He then expressed the formal equations graphically so that the analyst could simply and easily make three log readings, enter a chart, and exit with a value of water saturation, S_w. Fig. 10.1 illustrates that the chart used for this method.

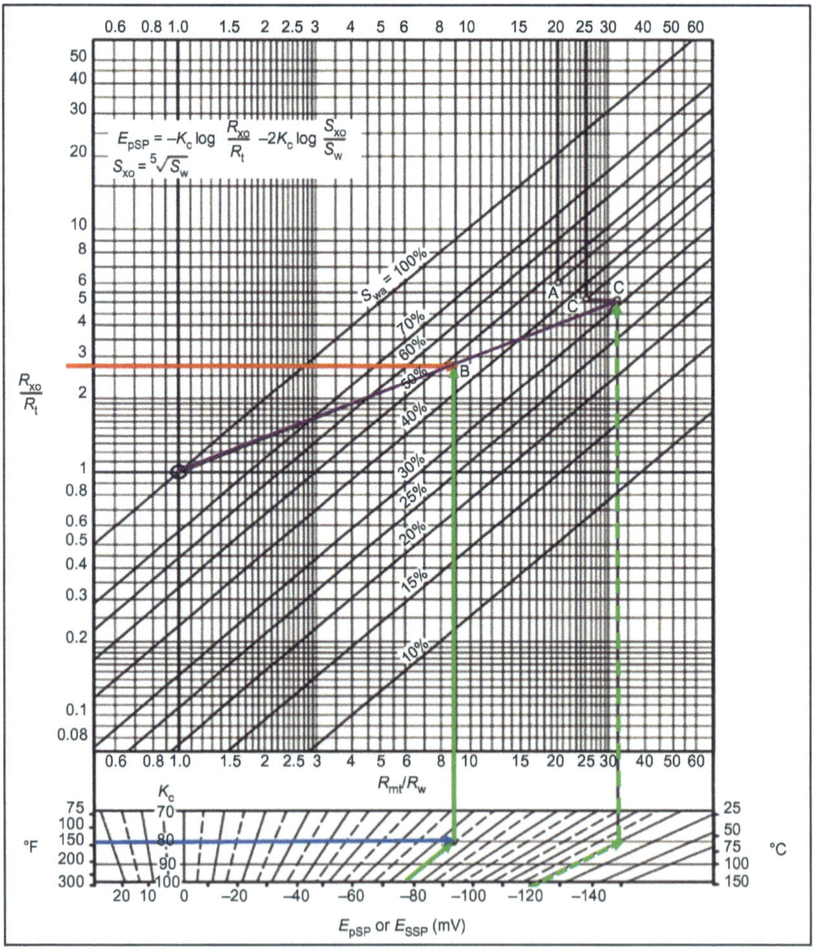

Figure 10.1 Chart for S_w from R_{xo}/R_t ratio and the SP deflection. *Image courtesy Schlumberger.*

This "Rocky Mountain method" is a perfect example of the ingenuity of the early log analysts. When faced with few "knowns" and many "unknowns" they resorted to a clever device to eliminate entirely what was not known by taking ratios. What was left was a useful answer product that solved many log analysis problems. Tixier published his paper in 1949 which showed that it was possible to deduce water saturation with the use of a resistivity/SP log alone without the need for a porosity log. To understand the simplicity and power of the method it is sufficient to write down Archie's equation for both the invaded and uninvaded zones:

$$S_w{}^2 = \frac{FR_w}{R_t} \text{ and } S_{xo}{}^2 = \frac{FR_{mf}}{R_{xo}}$$

Since porosity, and hence the formation factor F, are unknowns a ratio of these two equations eliminates the unknown quantity:

$$\frac{S_w{}^2}{S_{xo}{}^2} = \frac{R_w}{R_{mf}} \frac{R_{xo}}{R_t}$$

An assumption may be made about the relationship between invaded and uninvaded water saturations and, based on core flooding experiments and other observations, a general equation sets S_{xo} equal to $S_w{}^{0.2}$. The left-hand side of the equation now simplifies to:

$$S_w{}^{(8/5)} = \frac{R_w}{R_{mf}} \frac{R_{xo}}{R_t}$$

The final piece of the jigsaw puzzle is provided by the SP, which also depends on the ratio of the connate water resistivity, R_w, to the invading mud filtrate resistivity R_{mf}.

$$\mathrm{SP} = -K \log \frac{R_{mf}}{R_w}$$

From which we derive:

$$\frac{R_w}{R_{mf}} = 10^{(\mathrm{SP}/K)}$$

which leads to the final expression for S_w:

$$S_w = \left[10^{(\mathrm{SP}/K)} \frac{R_{xo}}{R_t}\right]^{8/5}$$

10.2 Application of the Rocky Mountain method

In practice, this equation was solved graphically with a plot of the R_{xo}/R_t ratio against the SP in mV as shown in Fig. 10.1. An example of the use of the chart is shown on the figure for a case where the SP is 78 mV (green line), the formation temperature is 150 °F (giving $K = 90$, see the blue line) and $R_{xo}/R_t = 2.8$ (red line). At the intersection of the red and green lines at point **B** the water saturation is read to be 48%.

An interesting wrinkle to this chart is its ability to make corrections for the effects of shale on the calculation of water saturation. In a shaly formation the deflection of the SP is reduced relative to its value in a similar formation that is shale free. This is covered in Chapter 9 in the discussion of the difference between the shale free maximum SP in a clean formation (SSP) and the reduced SP in a shaly formation (PSP). In Fig. 10.1 note that a second green line is drawn for an SP of 120 mV. If this represents a clean formation SSP and the 78 mV the PSP in a shaly formation, then the shale corrected vale of S_w is read at the intersection of the purple line and the second green line at point C as just 30%.

An alternative method to find S_w graphically is by using the Elgen nomogram as illustrated in Fig. 10.2 Note that this chart requires log

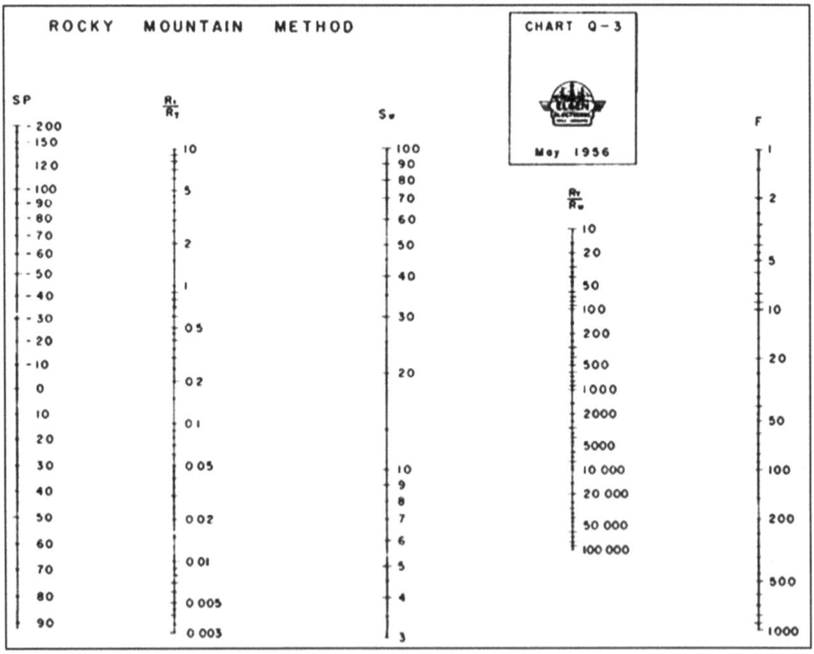

Figure 10.2 Nomographic solution for Rocky Mountain method (after Elgen)

readings of SP, in mV, and the ratio R_i/R_t, which is sourced from any available resistivity log. Chapter 3 gives more detailed instructions for deducing this ratio, either from an electric log, or laterolog, for R_t and, perhaps, from a micrpolog for R_i.

The Elgin chart takes an extra step in that once S_w has been found (by drawing a line from the SP value through the R_i/R_t value) the analyst may start afresh from the S_w value so determined and draw another line through the R_t/R_w column and extending it to the F column. This gives a "back door" approach to porosity. However, it is doubtful for the analyst to know the exact value of R_w, and although this right-hand part of the chart follows Archie's arithmetic, its usefulness is a little shaky. All the same it is an interesting historical comment on the lengths the early log analysts went to in order to find what they needed from very limited resources.

Exercise 10.1

(Use Elgen Chart)

$R_i = 20\ \Omega\text{m}$, $R_t = 10\ \Omega\text{m}$, SP = 90 mV, find S_w

If $R_t/R_w = 200$, find formation factor F

10.3 Effects of invasion

With the introduction of the induction-16″ Short Normal logs it was possible to redraw the basic Rocky Mountain chart to include the effect of mud filtrate invasion. Fig. 10.3 shows the version adapted for the Induction-SN combination. As before the SP and formation temperature are entered on the X-axis but on the Y-axis the input is the ratio of $R_{16''SN}/R_{ID}$. Note that the lines of constant S_w are now slightly curved and that all 100% water bearing points fall in a shaded area of the plot rather than on a single line. Note also that the S_w scaling is invasion dependent.

Exercise 10.2

(Use Fig. 10.3**)**

Given: Invasion is "moderate"

$R_{16''SN} = 20\ \Omega\text{m}$, $R_{ID} = 10\ \Omega\text{m}$, SP = 90 mV, and $T = 150°\text{F}$

Find S_w

With the introduction of an improved method for measuring the resistivity in the invaded zone (R_{xo}), the Rocky Mountain interpretation chart changed to keep pace. Fig. 10.4 shows the version that correspond for the induction/SFL combination.

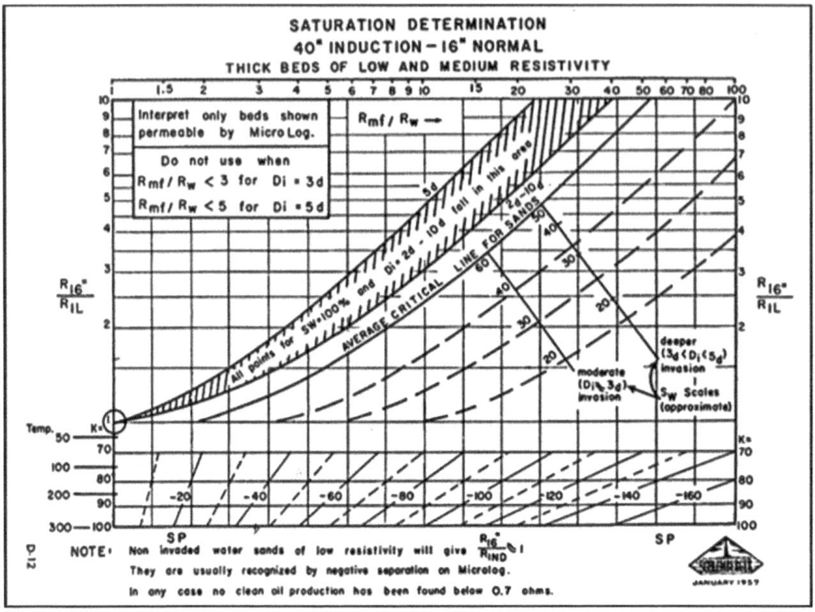

Figure 10.3 Graphical solution for S_w from Induction/16″ Short Normal log suite. *Image courtesy Schlumberger.*

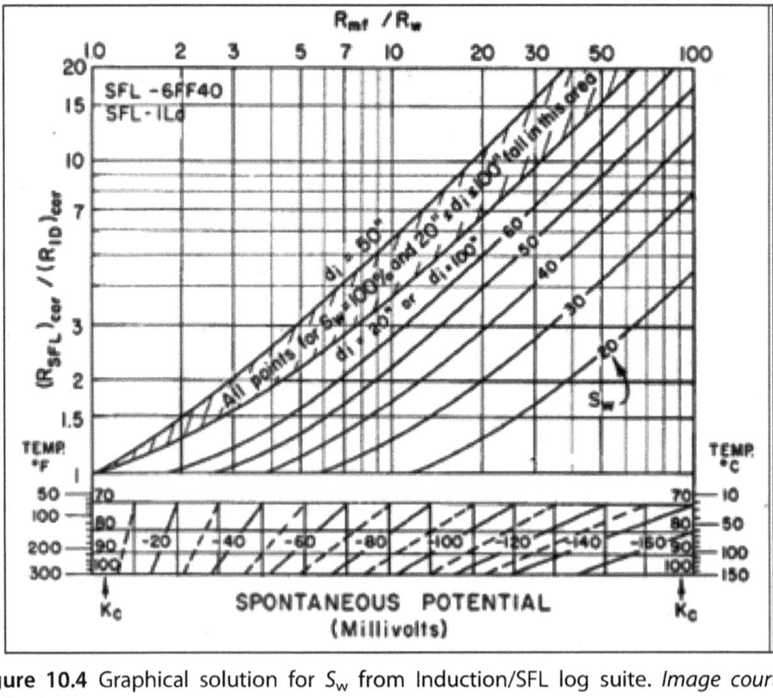

Figure 10.4 Graphical solution for S_w from Induction/SFL log suite. *Image courtesy Schlumberger.*

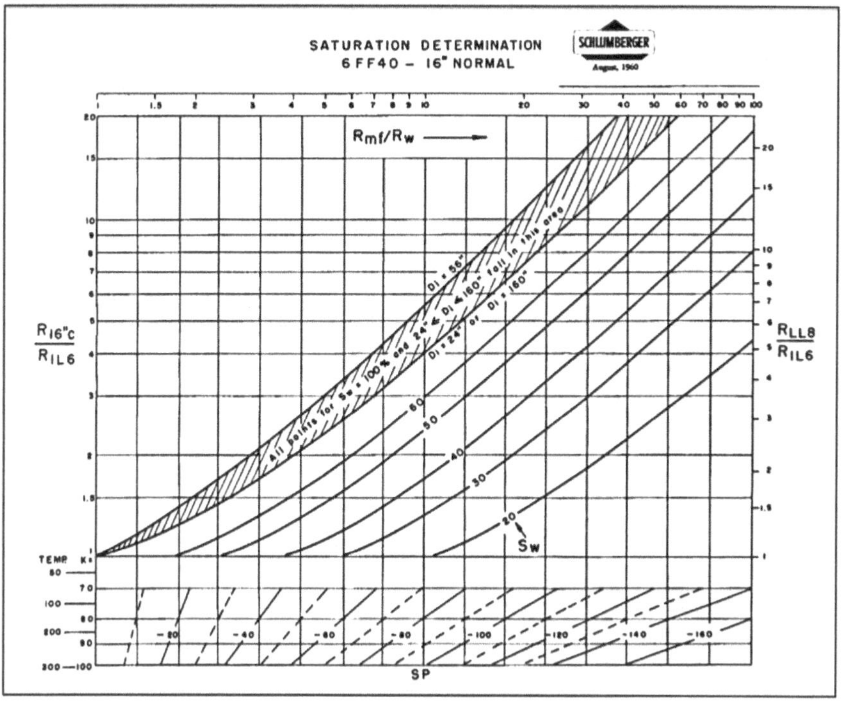

Figure 10.5 Graphical solution for S_w from 6FF40 induction/LL8 combination. *Image courtesy Schlumberger.*

Another combination arose with the introduction of the Dual Induction-Laterolog-8 combination. Fig. 10.5 gives a version of the Rocky Mountain chart with *Y*-axis scaled for both the 16″ Short Normal (on the left-hand side) and the Laterolog-8 (on the right-hand side).

10.4 Uses for quick look

While single depth levels in a logged well may be analyzed using the charts shown earlier, the introduction of the induction combination tool made it possible to perform some raw log measurement "manipulations" on the fly at the wellsite during the logging operations. This allowed, for example, the generation of an R_{wa} curve while logging an induction-sonic combination. Likewise the Rocky Mountain method could also be applied in a semi-continuous way by the generation of a scaled ratio of the R_{xo} to R_t curves that would overlay the SP in such a way that in water bearing porous sections and shales the two curves would coincide but in the presence of hydrocarbons the two curves would separate. An example of this kind of presentation is given in Chapter 13.

The lesson learned from the early log analysts, such as Maurice Tixier, is that when one takes a ratio of two numbers it is possible to eliminate unknowns and yet still arrive at an important result. In the case of the Rocky Mountain method the value of R_w, R_{mf}, and ϕ can be unknowns and yet, with just two resistivity measurements and an SP, the vital formation evaluation parameter of water saturation can be found.

Further reading

Tixier, M.P., 1949. Electric-log Analysis in the Rocky Mountains. American Petroleum Institute.

Exercise Solutions

Exercise 1: $S_w = 30\%$ and $F = 17$
Exercise 2: $R_{mf}/R_w = 14$ and $S_w = 40\%$

CHAPTER 11

Pickett and Hingle Plots

Abstract

This chapter covers the manual methods used with pre-digital well logs to determine connate water resistivity, R_w, and the rock matrix (zero-porosity) responses of the main "porosity" tools, namely the sonic, neutron, and density. Today these two parameters are derived from plots that are regularly run as subroutines in preinterpretation passes that automated interpretation packages offer. However, the analyst of yesteryear, working with paper prints, was obliged to build these plots manually using graph paper and laboriously reading pairs of log readings, plotting them, and then constructing the required trend lines that, hopefully, pinpointed the required parameter, which could then be used in the main calculations of formation porosity and water saturation.

A corner of the Twingon oil reserve. Burma—1910. Reproduced from "Oil-Finding," by E. H. Cunningham, Craig, Published by Longman, Green and Co., London (undated) (pp. 124-125)

11.1 Pickett plot basics

This plot, first proposed by G.R. Pickett, requires a resistivity log and a porosity log. The resistivity log can be an induction log or a laterolog or

Formation Evaluation with Pre-Digital Well Logs
DOI: https://doi.org/10.1016/B978-0-12-820232-6.00011-X

even, under favorable conditions, an electric log. The porosity log can be a sonic, density, or even a neutron log from which porosity has already been derived from knowledge of the zero-porosity response of the specific sensor involved. In the next section of this same chapter, under *Hingle Plots*, how zero-porosity responses are found will be addressed.

The Pickett plot requires that log readings be taken in a 100% water-bearing section where there is a range of porosities. These conditions may not always be met and so the method, though theoretically sound, does have its limitations. The plot itself is made on log–log graph paper with the values of R_t plotted on the Y-axis and values of porosity on the X-axis. The standard Pickett plot graph paper sheet will have five decades on the Y-axis and two decades on the X-axis. It is normal to plot the porosity as a fraction of one. Thus the Y-axis is scaled from 0.01 to 1000, and the X-axis is scaled from 0.01 to 1.

The basis of the plot is a simple graphical representation of Archie's equation:

$$S_w{}^n = \frac{aR_w}{\phi^m R_t}$$

Taking logs of both sides of the equation and rearranging the terms we can derive an expression for log R_t:

$$\log R_t = -m\log\phi + \log R_w + \log a - n\log S_w$$

which is of the form

$$y = mx + c$$

When a series of paired resistivity and porosity readings from the same reservoir unit are plotted on a log R_t versus log ϕ graph paper the 100% water-bearing points will lie on a straight line with a slope of $-m$. An extrapolation through these water-bearing points to the 100% porosity line will indicate the value of the aR_w product. The appendix to this chapter shows a full-page blank sheet of log–log graph paper (Figure A11.1) for the reader's use, should the need arise.

To illustrate the method, refer to Fig. 11.1 where points from a reservoir with porosities varying from 8% to 32% and water saturations from 100% on down are plotted. Note that the cloud of points has a distinct limiting "edge" to the south west of the plot through which a green line has been drawn. This line serves two purposes. Its slope is equal to $-m$ and its extrapolation to the Y-axis intersects at aR_w. The reader is invited to numerically determine the parameters m and aR_w. A suggested answer is available at the end of the chapter.

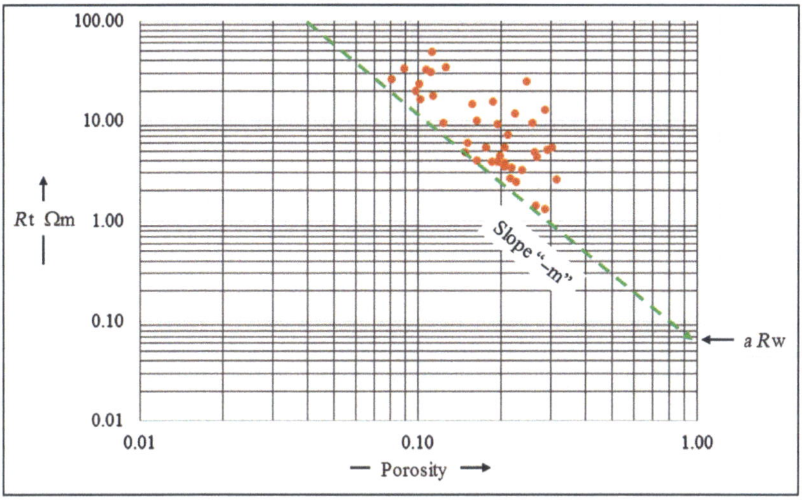

Figure 11.1 An example Pickett plot for $-m$ and aR_w.

11.2 Reading Pickett plots for "*m*" and "aR_w"

Once the raw log readings have been posted on the log–log graph paper used for Pickett plots the job still remains to determine the exact values of both the Archie "*m*" factor and the "aR_w" number. The slope of a line on log–log paper is found by marking any two points on the line and noting the *X* and *Y* values for each. It is then a matter of simple arithmetic, as shown in Fig. 11.2 and by Table 11.1.

11.3 Reading Pickett plots for S_w

On a Pickett plot all the points that fall on the dashed green line (Figs. 11.1 and 11.2) are at 100% water saturation. For any chosen porosity value, as read on the *X*-axis, the resistivity on that line corresponds, by definition, to R_o for that particular porosity. At the same porosity any point *above* the dashed green line will be at water saturation *less* than 100%. If Archie's equation is written as

$$R_t = \frac{R_o}{S_w{}^n}$$

and *n* is defaulted to 2, then we can find the value for R_t for any chosen value of S_w in terms of a multiple of R_o.

If $S_w = 0.5$, then $R_t = R_o/(0.5)^2 = R_o/0.25 = 4\ R_o$.
If $S_w = 0.25$, then $R_t = R_o/(0.25)^2 = R_o/0.0625 = 16\ R_o$.
If $S_w = 0.1$, then $R_t = R_o/(0.1)^2 = R_o/0.01 = 100\ R_o$.

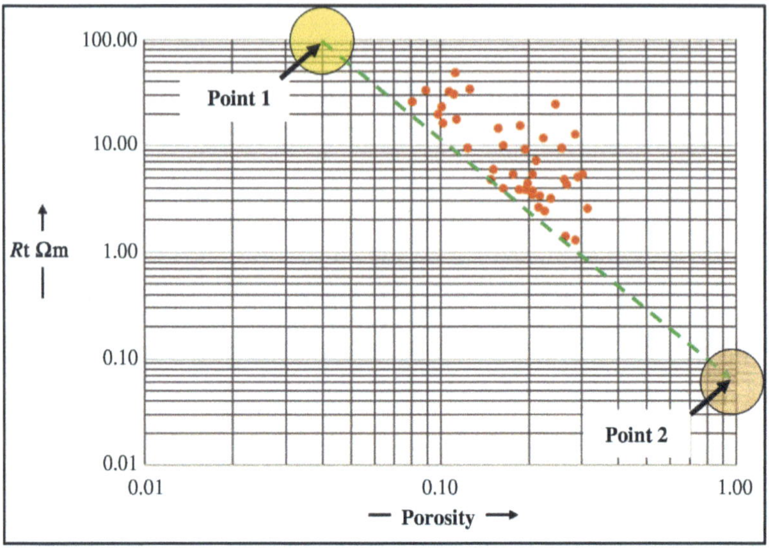

Figure 11.2 Pickett plot analysis for "*m*" and "aR_w"

Table 11.1 Finding "*m*" from Pickett plot.

Point	1	2
X Value	ϕ_1	ϕ_2
Log (X Value)	a	b
Y Value	R_1	R_2
Log (Y Value)	c	d
Δ Log Y	c-d	
Δ Log X	a-b	
Slope	(c-d)/(a-b)	

This can be made visual on a Pickett plot by the simple expedient of drawing a line parallel to the 100% S_w line passing through the corresponding multiple of R_o. For this it is convenient to pick a point on the green dashed line that passes through, for example, R_t exactly equal to 1 Ωm, which in this case occurs at a porosity of 30%. The 50% S_w line will then have to pass through the 30% porosity at 4 Ωm (4 R_o) and so on The completed set of equi-saturation lines is shown in Fig. 11.3.

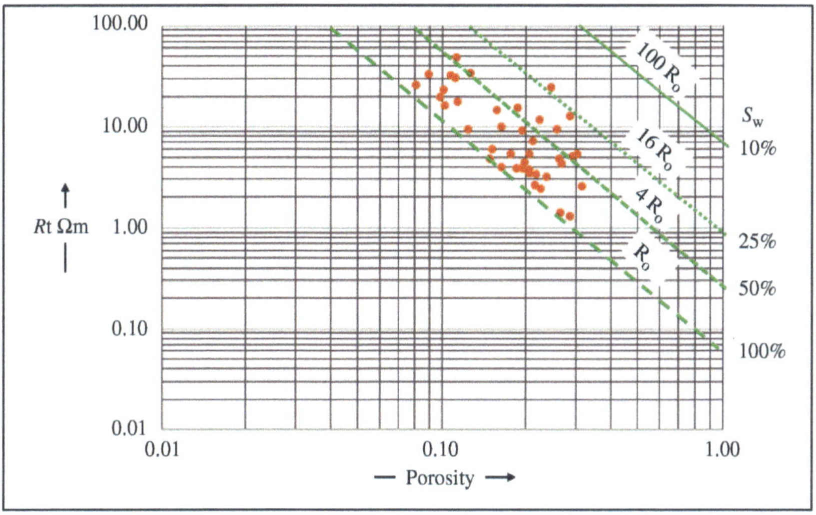

Figure 11.3 Water saturation from Pickett plot.

From the plot of Fig. 11.3, the distribution of S_w can be seen. Half a dozen points lie between 25% and 50% water saturation. The remaining points depict a saturation profile through a transition zone to the few points that actually define the 100 S_w line.

11.4 Hingle plots basics

The Hingle plot is used to pin down the zero-porosity response whatever porosity sensitive tool is available and used. This could be either a sonic, density or neutron tool, and the parameters sought are, respectively, Δt_{ma}, ρ_{ma}, or ϕN_{ma} (which is usually close to 0). The means to this end is a plot of the raw porosity tool measurement paired with a function of R_t, as derived from the available resistivity tool (induction, laterolog, or electric log). A generic Hingle plot is shown in Fig. 11.4. Note that water-bearing points extrapolate to the zero-porosity matrix point (ρ_{ma}, ϕN_{ma}, or Δt_{ma}).

The function of R_t used depends on the parameters used in the Archie saturation equation. This is most easily understood by considering the standard Archie equation and the Hingle rearrangement of the terms involved:

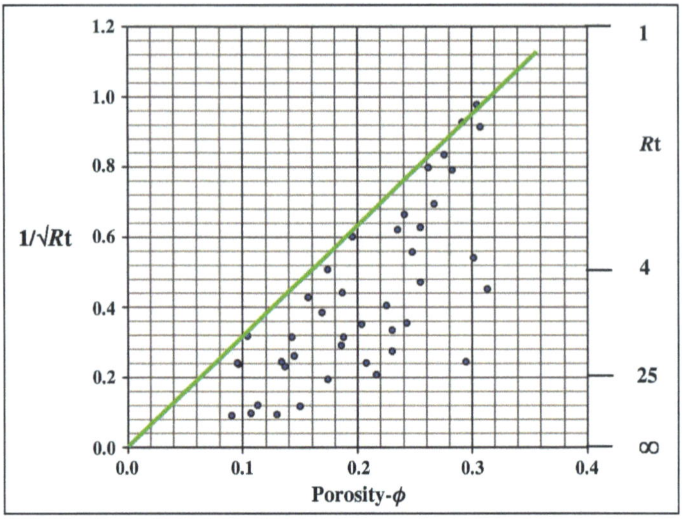

Figure 11.4 Linear $1/\sqrt{R_t}$ function versus porosity plot scaling.

$$\text{Standard Archie equation:}\ \ S_w^n = \frac{a}{\phi^m}\frac{R_w}{R_t}$$

In "hard rock" (lower porosity carbonates for example) we can set $a = 1$ and $m = 2$, leading to the Hingle rearrangement for the relation between porosity and resistivity in 100% water-bearing points (when S_w is set to 1) as

$$\phi = \left(\frac{R_w}{R_t}\right)^{\frac{1}{2}}$$

Thus a plot of porosity against the reciprocal square root of R_t, $\left(\sqrt[2]{\frac{1}{R_t}}\right)$, abbreviated RRRT, will show points lying on a straight line. On such a scale, zero porosity will yield an infinite resistivity, the reciprocal root of which will be zero. A straight line through the water-bearing points will cross the X-axis at the zero-porosity point, which can be related back to the sought-after value for Δt_{ma}, ρ_{ma}, or ϕN_{ma}.

Likewise, in less consolidated formations (sand/shale sequences) we can set $a = 0.62$ and $m = 2.15$, leading to

$$\phi = \left(\frac{0.62\ R_w}{R_t}\right)^{\frac{1}{2.15}}$$

And thus the resistivity function changes to $\left(\sqrt[2.15]{\frac{0.62}{R_t}}\right)$.

These relationship give rise to the resistivity (Y-axis) scaling on the Hingle plots shown in Fig. 11.4. Essentially the plot is a linear function of resistivity versus a linear porosity. Note that there are two Y-axis scales. On the left of the plot is the linear scale for the $1/\sqrt{R_t}$. On the right are the nonlinear actual values of R_t. The actual resistivity runs from ∞ to 1, decreasing up the page, while the RRRT scale *increases* up the page from 0 to 1. The porosity scale increases from 0 on the left to 0.4 (or 40%) on the right. In practice, the porosity is yet to be known so on this axis ϕ_N and Δt will be plotted increasing to the right and ρ_B increasing to the left. That way density porosity, ϕ_D, will increase to the right. In practice, the analysts may use one or the other of the full sized Hingle plot graphing papers reproduced in the Appendix as Figures A11.2, for hard rock and Figure A11.3 for soft rock.

11.5 Hingle plots for R_w and S_w

While the principal purpose of the Hingle plot is to find the value of the zero-porosity point for a sonic, neutron, or density logging device run in a formation with an unknown rock matrix a side benefit is the addition of a "quick and dirty" visual analysis for the water saturation (S_w) of each plotted point and a simple way to find the connate water resistivity, R_w as

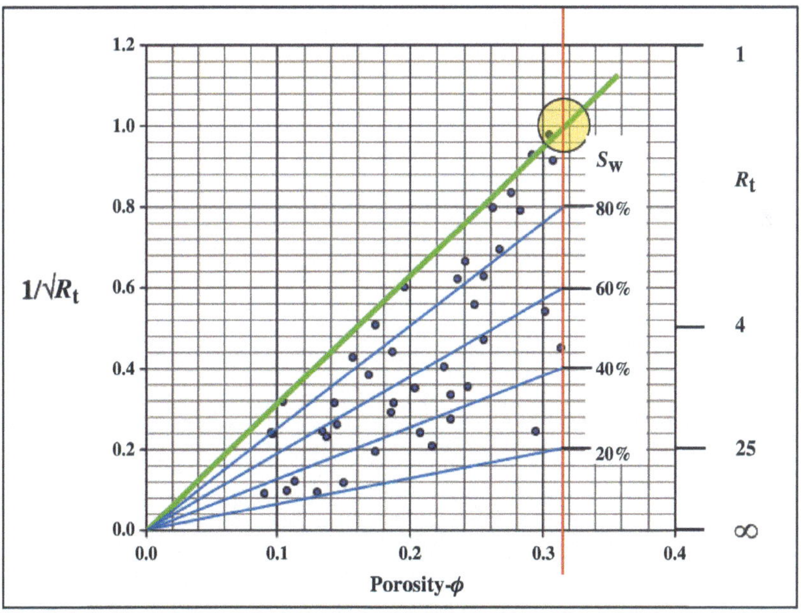

Figure 11.5 Finding R_w and S_w from a Hingle plot.

well. Fig. 11.5 illustrates how both R_w and S_w can be found from a generic Hingle plot.

For R_w determination all that is needed is to seek the porosity that corresponds to the 100% S_w where R_t (and $1/\sqrt{R_t}$) $=1$. On Fig. 11.5, the green line through the most north westerly points at a porosity of 31.5%. Archie's equation can then be applied in reverse to yield:

$$R_w = R_t\,\phi^m$$

Since R_t is, by choice, 1 Ωm and m is 2.24 (as determined earlier from the Pickett plot) we find

$$R_w = 0.315^{2.24} = 0.075$$

The determination of S_w for each point is equally straight forward. The red line on Fig. 11.5 is drawn vertically down the page from the point where the green line intersects the R_t value of 1. A "fan" of blue lines delineate constant values of S_w for 80%, 60%, 40%, and 20%. In this data set there are 10 levels plotted, which have S_w between 20% and 40%.

A data set with raw sonic log data is shown in Fig. 11.6. The reader is encouraged to analyze the plot and determine the value of Δt_{ma} and R_w assuming that Δt_{Fluid} is 189 μseconds/ft. The green line will help in determining Δt_{ma} and the dashed red line the porosity when $R_t = 1$ Ωm, which then leads to R_w. A suggested solution is given in the Appendix to this chapter.

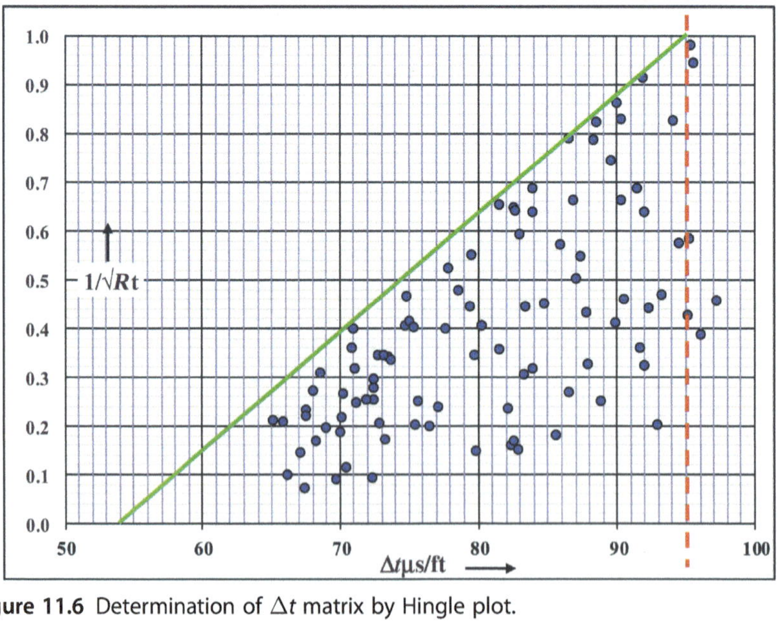

Figure 11.6 Determination of Δt matrix by Hingle plot.

11.6 Teaming Pickett and Hingle methods

In many ways the two plots discussed in this chapter are complimentary. With the Pickett plot the analysts can determine the Archie "*m*" factor and the value of the "aR_w" product given that the porosity is known. In many cases the porosity is not known a priori because the rock type has yet to be determined and so the zero-porosity matrix values for the porosity sensors are still unknowns. It may come down to a "chicken and egg" situation where an iteration is required to completely tie down the calculated porosities and hence water saturations. For "first pass" scouting for promising zones the analysts is fully justified in making initial guesses as to the rock type and for working on the assumption that "*a*" = 1 and "*m*" = "*n*" = 2. The refinements can come later when the presence of economically recoverable hydrocarbons is clearly established.

Appendix

Solution to Pickett plot exercise

Any two points on the 100% S_w line are picked. For ease of reading the numbers in this particular case are selected as:

Point 1: 4% porosity with 100 Ωm resistivity (0.04,100) and

Point 2: 100% porosity with 0.075 Ωm resistivity (1, 0.075).

Using the table, the slope is found to be −2.24 indicating that *m* is +2.24. The 100% S_w line intersects the *Y*-axis when it reaches 100% porosity at the aR_w value of 0.075 Ωm.

Point	1	2
X Value	0.04	1
Log (X Value)	-1.39794	0
Y Value	100	0.075
Log (Y Value)	2	-1.12494
Δ Log Y	3.12	
Δ Log X	-1.40	
Slope	-2.24	

Application of these results calls for calculations of S_w using:

$$S_w = \sqrt[n]{\frac{0.075}{\phi^{2.24} R_t}}$$

Solution to Hingle plot exercise

In Fig. 11.6, the green line through the 100% water-bearing points intersects the Δt axis at 54 μseconds/ft, indicating a limy sandstone matrix.

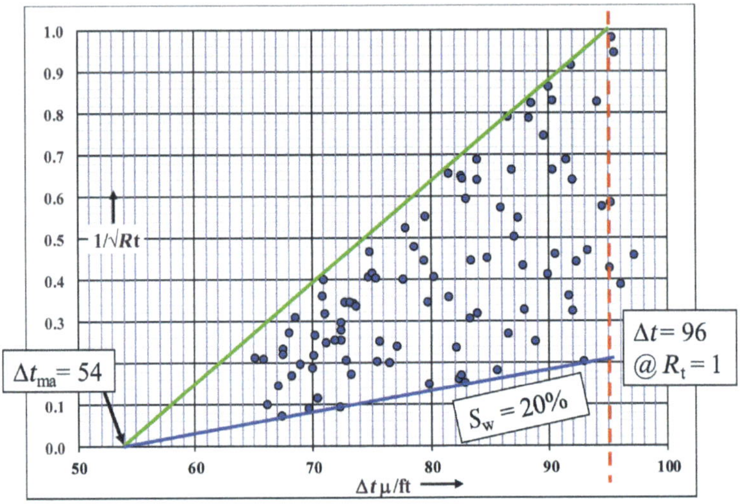

The dashed red line intersects the $R_t = 1$ line when drawn from a Δt of 96 μseconds/ft. The sonic porosity at this point is given by

$$\phi_S = \frac{\Delta t - \Delta t_{ma}}{\Delta t_{Fluid} - \Delta t_{ma}} = \frac{96 - 54}{189 - 54} = \frac{42}{135} = 0.311$$

R_w is thus given by $0.311^2 = 0.097\ \Omega m$ (assuming $m = 2$). The lowest S_w is 20%.

Further reading

Hingle, A.T., 1959. The use of logs in exploration problems. SEG 29th Annual Meeting.

Lindley, R.H., 1961. The Use of Differential Sonic- Resistivity Plots to Find Movable Oil in Permian Formations. Society of Petroleum Engineers.

Pickett, G.R., 1966. A review of current techniques for determination of water saturation from logs. Jour. Pet. Tech. 18 (I), 1425–1435.

Pickett, G.R., 1973. Pattern recognition as a means of formation evaluation. Log Analyst 14 (4), 3–11.

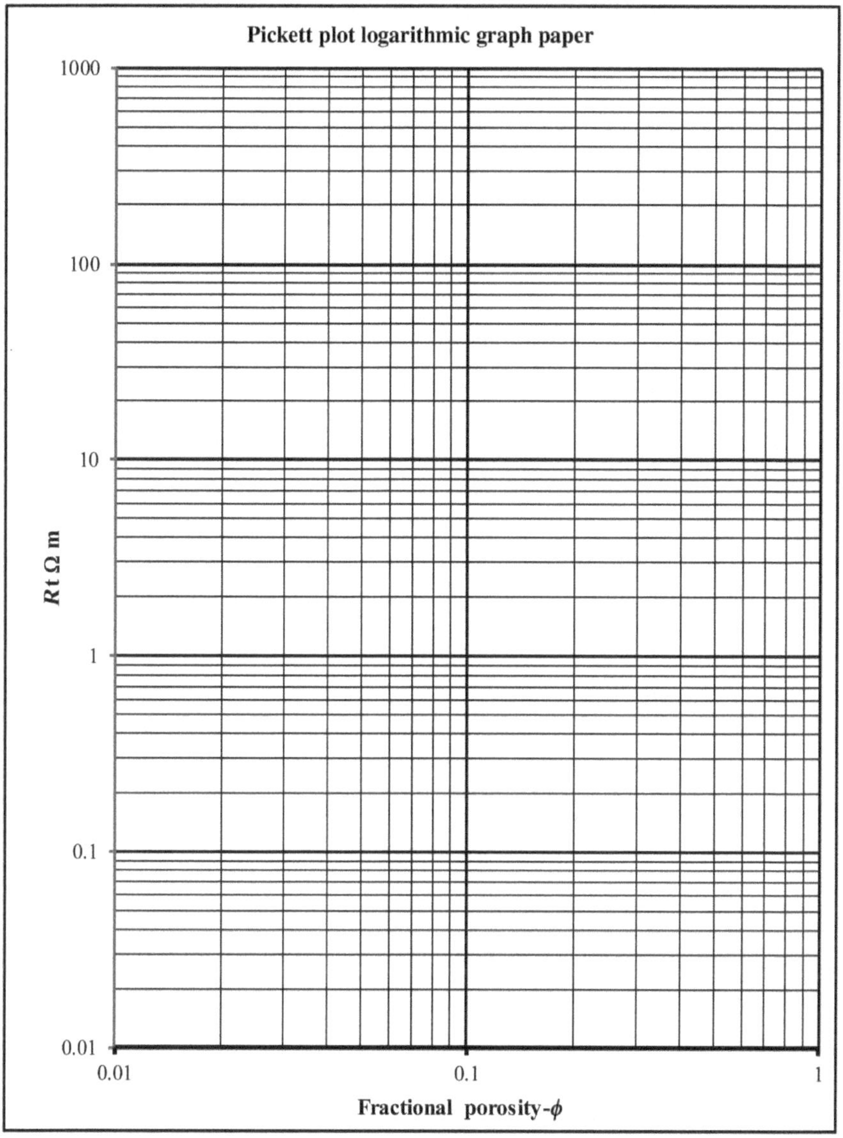

Figure A11.1 Pickett plot graph paper.

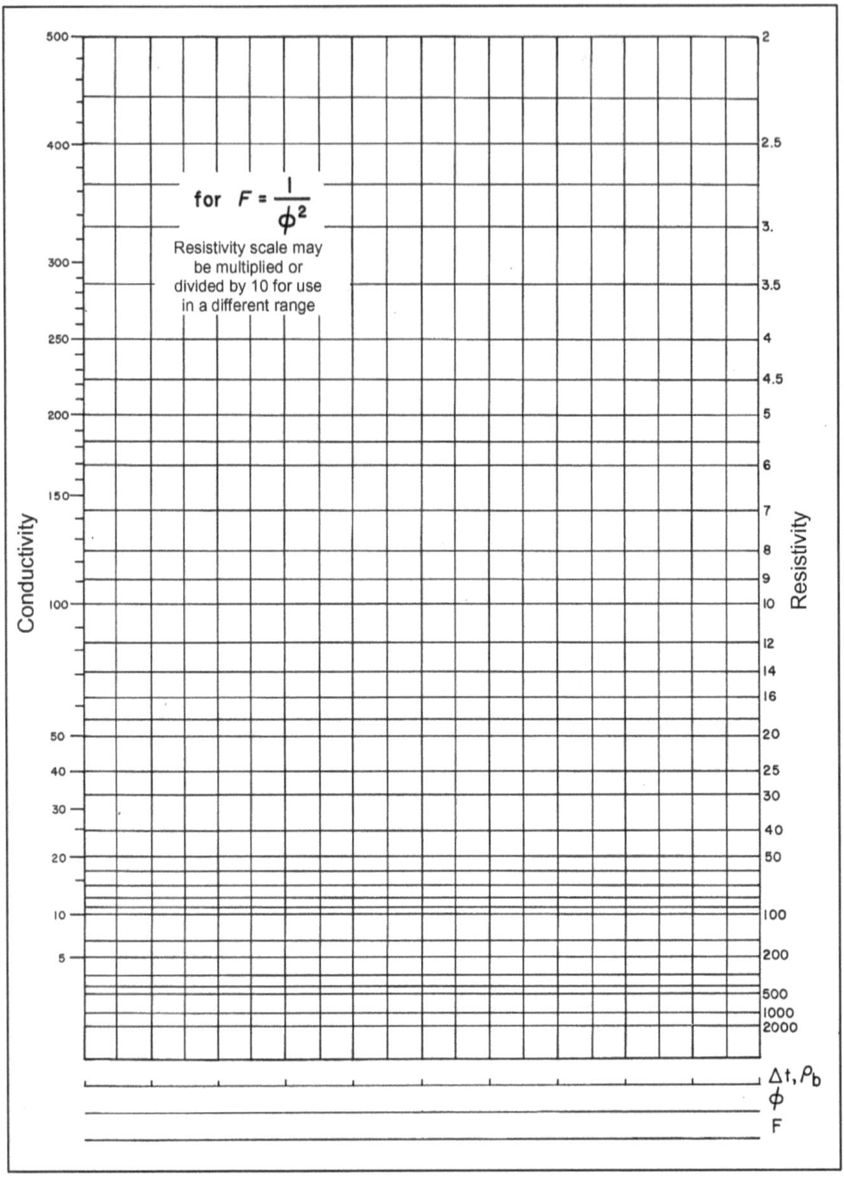

Figure A11.2 Hingle plot scaling for soft rock resistivity scales. *Image courtesy Schlumberger.*

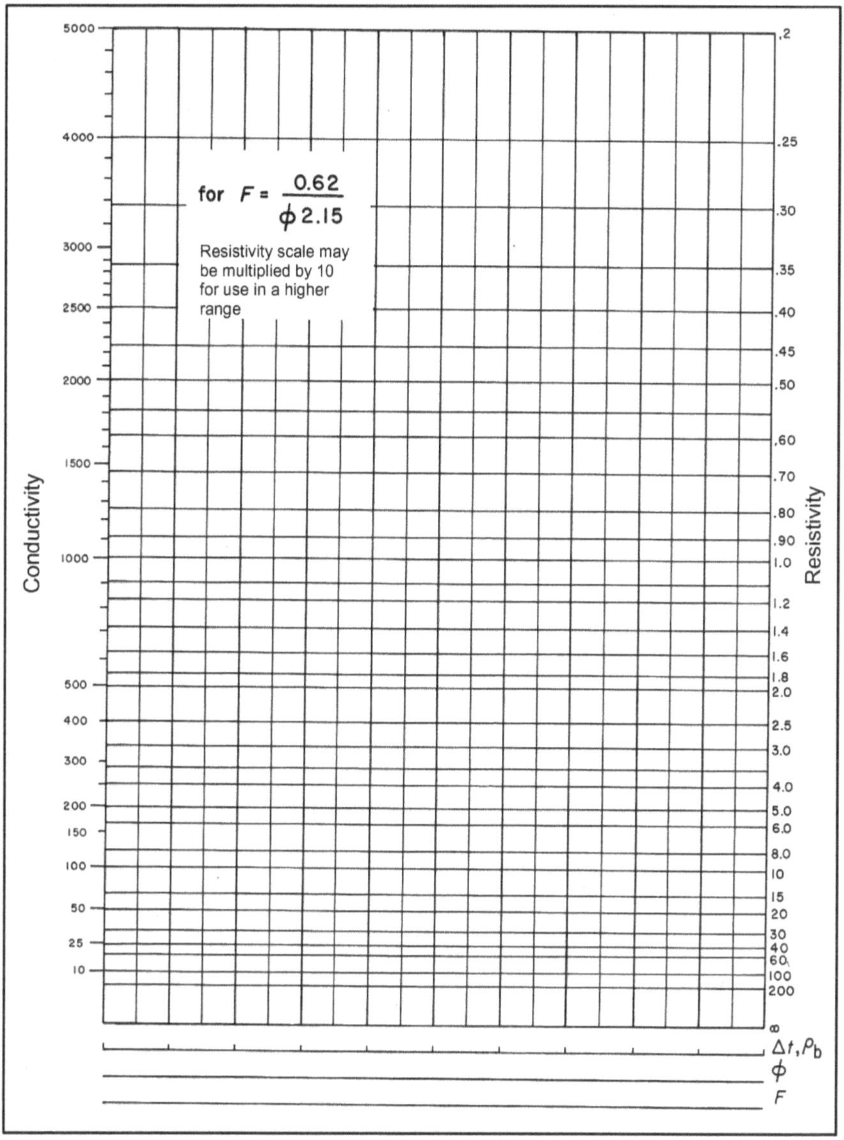

Figure A11.3 Hingle plot scaling for hard rock resistivity scales. *Image courtesy Schlumberger.*

CHAPTER 12

Logging Empty Holes

Abstract

In this chapter, the special case of running and interpreting logs recorded in empty, that is, air-drilled, holes is discussed. Practical guidelines are given for quantitative analysis using the pre-digital logs that were available at the time these analysis techniques were developed and put into practice.

Dad Joiner congratulates Doc Lloyd after the drill stem test on the Daisy Bradford well. H. L. Hunt (with cigar) is third from the right. Soon afterward, Hunt bought Joiner out, a deal which was to be the cornerstone of the Hunt fortune. Courtesy the Barker History Centre, Austin, Texas.

12.1 Why drill empty holes?

There are a number of reasons for drilling and logging "empty holes." Chief among them is the avoidance of contact between an aqueous drilling mud filtrate and swelling clays that may be found in the target formation. Thus, in the absence of drilling mud to bring cutting to the surface,

Formation Evaluation with Pre-Digital Well Logs
DOI: https://doi.org/10.1016/B978-0-12-820232-6.00012-1

the driller uses a high volume, high speed, air compressor to literally "blow" the cuttings back up the anulus to surface. One of the consequences of this drilling technique is that when a gas bearing formation is encountered, combustible gas quickly arrives to surface. Thus, for safety reasons, the drilling is conducted through a multi-phase surface separator that allows the solids, liquids, and gases to be handled properly. Normally a flare pipe will be setup with a pilot burning all the time so that when gas is struck it will exit from the flare pipe and burn on the spot. Driving along a lease road close to where empty-hole drilling is taking place can be quite spectacular as great jets of flame erupt from alongside a drilling rig from time to time.

One of the consequences of empty-hole drilling is that some conventional logs do not work. The log analysts must therefore use some interesting analysis tools in order to properly compute porosities and saturations. We will review two of such methods in this chapter and illustrate them with reference to some old logs. As a start we need to know which logs will work in empty holes and which ones will not.

Electrical type resistivity devices need a conductive mud in order to make their measurements of R_t, so in empty holes where there is no aqueous liquid mud the laterolog and microlog tools will not work. In such a case only the *induction log* will work. Likewise, many of the porosity logging devices such as the sonic and/or the thermal neutron tools require a liquid-filled hole for proper functioning. As a result:

no compensated neutron log,
use epithermal neutron pad contact type tool instead or
use old neutron-gamma, and/or
use gamma–gamma formation density tool—a pad contact tool which will work fine provided the hole is not rugose.

12.2 Log analysis model for standard (liquid-filled) holes

Fig. 12.1 shows a schematic for standard "Archie" type log analysis. The unit is divided into matrix, occupying $1 - \phi$, and fluid occupying ϕ. The porosity, in turn, is subdivided into liquid and gas, and the liquid into oil and water. The corresponding conventional log analysis symbols are:

S_w for water saturation
S_o for oil saturation
S_g for gas saturation

Conventional analysis for clean formation derives ϕ from neutron and density logs and R_t from either induction or laterolog devices and solves for water saturation, S_w using Archie's equation.

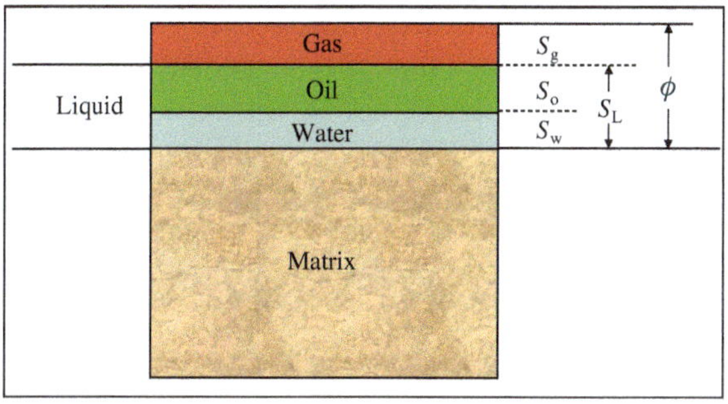

Figure 12.1 The conventional log analysis model.

12.3 Log analysis model for empty holes

Fig. 12.2 shows a schematic diagram for log analysis in empty holes. Note that now, as before, the matrix fraction of the unit volume is $1 - \phi$, but the porosity, ϕ, is subdivided into gas saturation, S_g, and liquid saturation, S_L. Interpretation in this case relies on the observation that since any near wellbore pore space will be filled with near atmospheric pressure air or gas then both the hydrogen index and density of the gas (or air) will be effectively zero. Thus, the neutron porosity, ϕ_N, is effectively equal to S_L. The density log will see the formation bulk density as

$$\rho_B = \rho_{Matrix} * (1 - \phi) + \rho_{Liquid} * \phi$$

from which the porosity may be derived as

$$\phi = \frac{\rho_{ma} - \rho_B}{\rho_{ma} - \rho_{Liquid}}$$

The matrix density may be obtained from local knowledge of the formation logged. For example, if the rock is a sandstone then the matrix

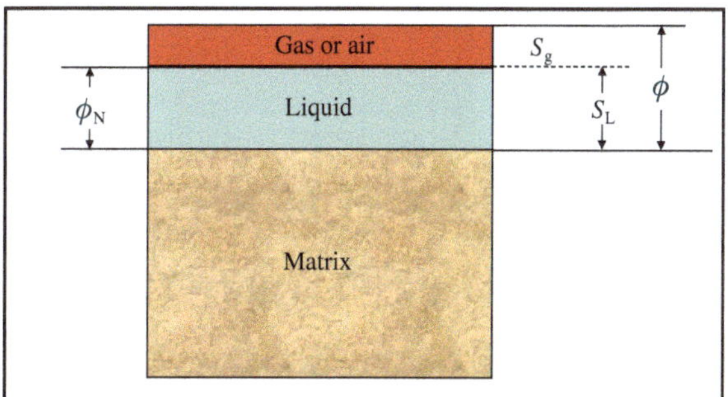

	Matrix	Liquid	Gas
Density	ρ_{ma}	1	0
Hydrogen index	0	1	0

Figure 12.2 Empty hole log analysis model.

density may be set to 2.65 g/cc. The remaining input for the calculation of porosity is the fluid density. This parameter depends on the liquid and gas densities and the liquid saturation. In the current case the liquid will be water with a density of 1 and the gas will be air (or perhaps gas) at near atmospheric pressure and have an effective density of 0, thus

$$\rho_{\text{Fluid}} = S_{\text{Liquid}} * 1 + S_{\text{Gas}} \times 0$$

$$\text{hence: } \rho_{\text{Fluid}} = S_{\text{Liquid}}$$

$$\text{therefore: } S_{\text{Liquid}} = \phi_{\text{N}}/\phi$$

then, with a little substitution, the formation porosity is derived as

$$\phi = \frac{\rho_{\text{ma}} - \rho_{\text{B}} + \phi_{\text{N}}}{\rho_{\text{ma}}}$$

This log analysis procedure for logs run in an empty hole may be illustrated by an example from the Clinton Sanstone. Fig. 12.3 shows

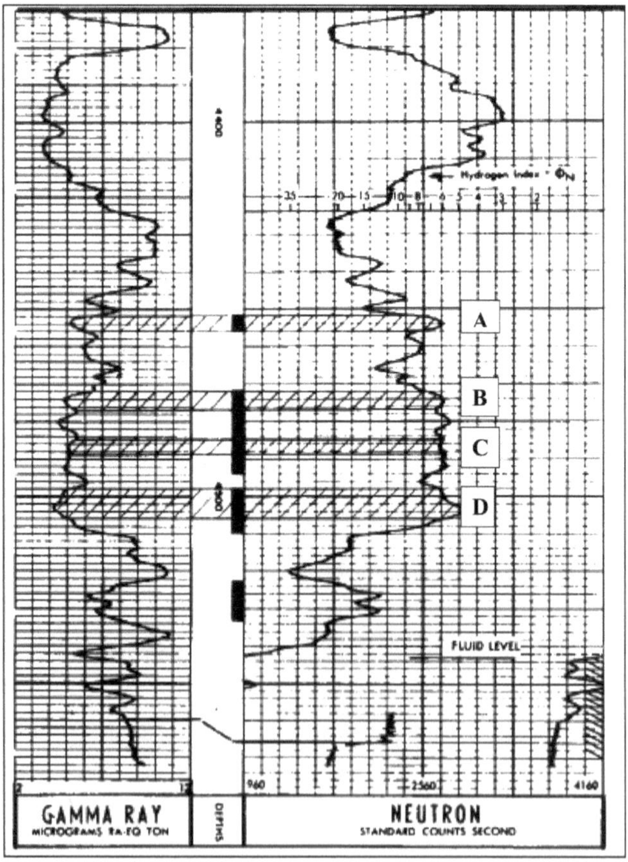

Figure 12.3 Clinton Sandstone Neutron Log.

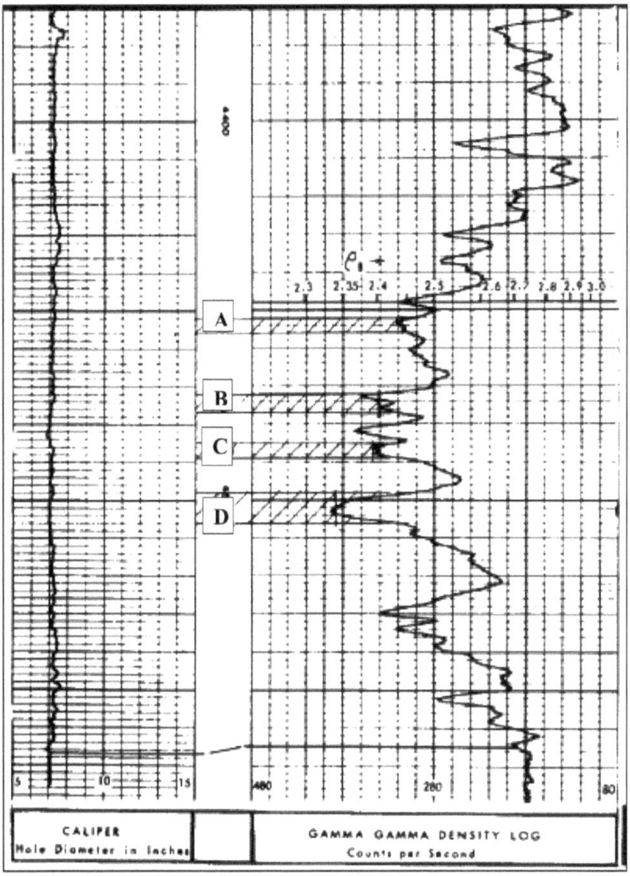

Figure 12.4 Clinton Sandstone Density Log.

the Neutron log and Fig. 12.4 shows the density log. Note that on both logs there are four zones marked as A, B, C, and D. Both logs are recorded in count rates. In other chapters the conversion of neutron counts to porosity and density countrs to density (g/cc) are fully covered. For the purposes of explanation of the empty-hole log analysis logic, and its practical application, the conversions from counts to porosity and density have been made already and are marked on the two logs. Fig. 12.3 has a scale in "hydrogen index" and Fig. 12.4 a scale in "ρ_B."

The reader is encouraged to solve for porosity and gas saturation in all four zones using the equations just derived and a calculator. A formaton matrix density of 2.68 g/cc should be used and the gas saturation, S_g, set equal to $1 - S_L$. Table 12.1 provides a convenient format to record the results of the calculations.

Table 12.1 Practical Log Analysis with Sandstone Example.

Zone	ρ_B gm/c	ϕ_N %	ϕ %	S_g %
A	2.44	6.0		
B	2.41	6.0		
C	2.40	6.0		
D	2.34	4.5		

The reader may check the results obtained with the "answer" table, which is provided at the end of this chapter. Note that this method does not require any resistivity data and provides interpreted values for porosity and gas saturation. What is not gas is considered "liquid" without any discrimination between oil and water.

12.4 Nomographic solution for log analysis in empty holes

As an alternative to using a calculator to derive prosoity and gas saturation from the neutron and density logs in the current example, a chart may be used instead. Such a chart is shown in Fig. 12.5. Nomograms[1] were commony used in the pre-digital logging era and can still be used today. The input at the bottom of the chart is the value of the formation bulk density, ρ_B. In this case the log reading from Zone A of the Clinton sandstone well is used and the staring point is 2.44 g/cc. On the *Y*-axis of the lower portion of the chart are values for "grain density"or, in our nomenclature, ρ_{Matrix}. In this case, core analysis has shown that the correct value to use is 2.68 g/cc. The red line is projected up and to the left to the intersection with 2.68 matrix density line and from there is extended vertically on the chart to the intersection with the netron porosity that is scaled on the *Y*-axis of the upper part of the chart. From the point of intersection of the red line with the blue line (correspondng to a ϕ_N of 6%) two green lines follow the "fan" to read S_{gas} to the right and slightly downwards (46%) and porosity (11%) going upward.

[1] Although the art of compiling nomograms to solve algebraic expressions is no longer taught a relevant text for the interested reader is listed at the end of this chapter.

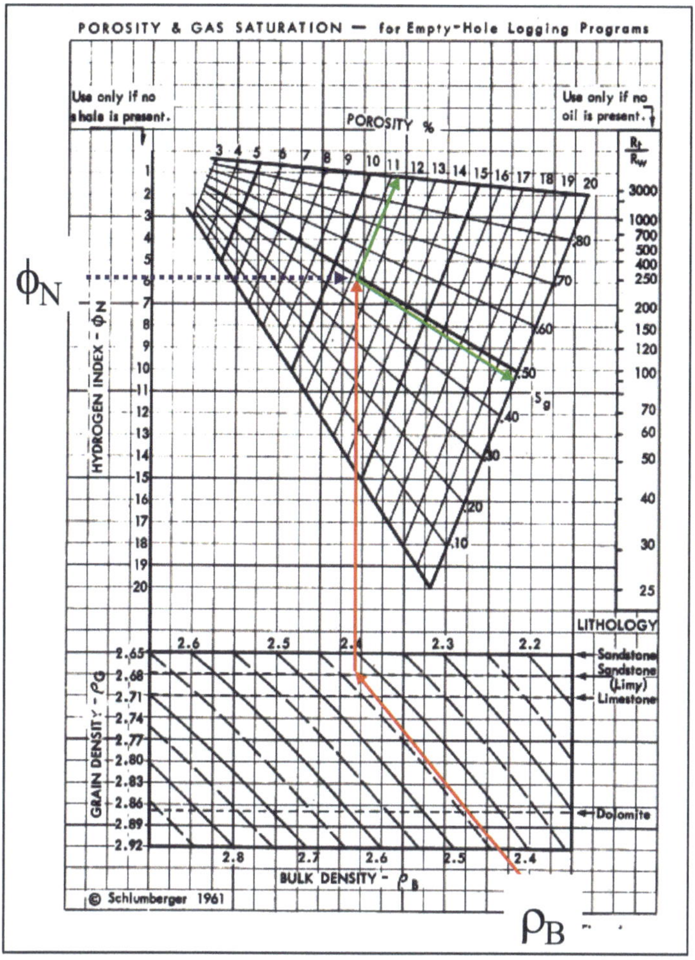

Figure 12.5 Nomographic chart for gas saturation and porosity. Image courtesy Schlumberger.

12.5 Alternative log analysis model for empty holes

Where resistivity logs are also available there is an alternaive method for log analysis that will yield both porosity and gas saturation. The starting point is a simplified Archie water saturation equation written in the form

$$S_w = \frac{1}{\phi}\sqrt{R_w/R_t}$$

We may also set

$$\rho_{\text{Liquid}} = S_w$$

When the porosity from density equation is written, S_w may be substituted for ρ_{Liquid} with the result that porosity can be defined in terms of formation density and resistivty as follows:

$$\phi = \frac{\rho_{\text{ma}} - \rho_{\text{B}}}{\rho_{\text{ma}} - \rho_{\text{Liquid}}} = \frac{\rho_{\text{ma}} - \rho_{\text{B}}}{\rho_{\text{ma}} - \frac{1}{\phi}\sqrt{\frac{R_w}{R_t}}}$$

Then, by simple substitution, we have

$$\phi = \frac{\rho_{\text{ma}} - \rho_{\text{B}} + \sqrt{R_w/R_t}}{\rho_{\text{ma}}}$$

Log analysis may be made by inputting values for bulk density and formation resistivity to this equation or by use of the nomographic chart shown in Fig. 12.5.

The method and its use can be illustrated by reference to the resistivity and density logs shown in Fig. 12.6 that were recorded on the Injun

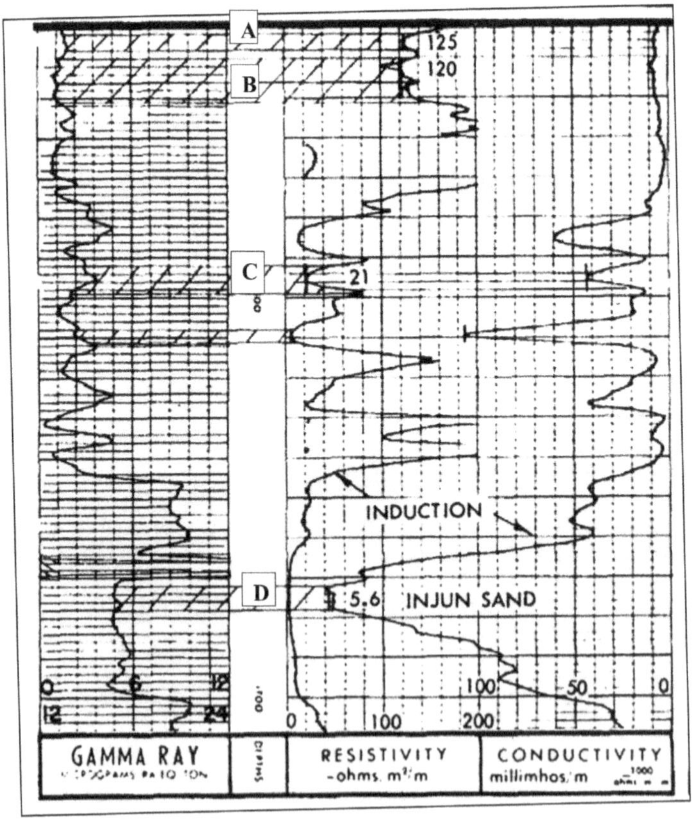

Figure 12.6 Resistivity log for the Injun sandstone example.

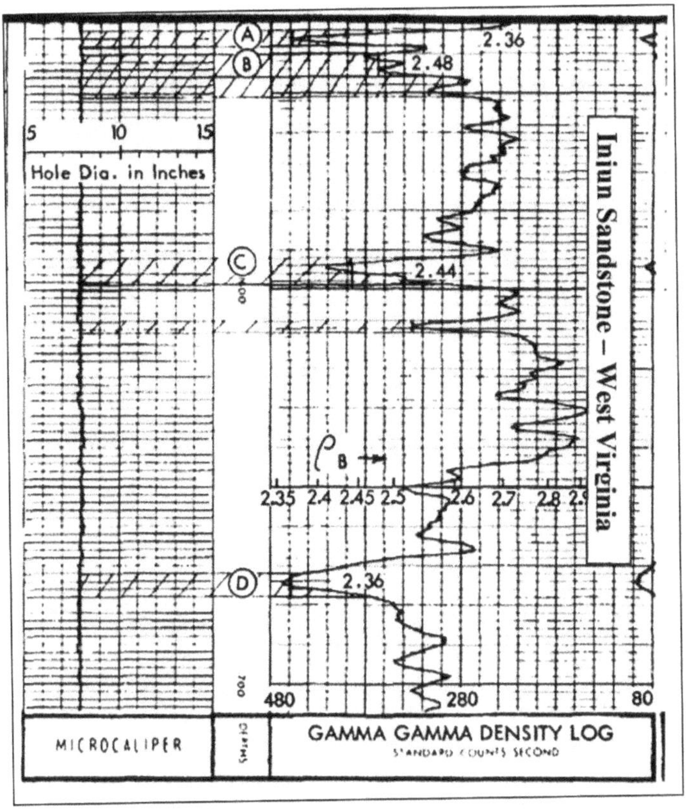

Figure 12.7 Gamma – gamma density log for the Injun sandstone example.

Sandstone from West Virginia. Again there are four zones marked on the logs A, B, C, and D (Fig. 12.7).

As an exercise the reader should work through the numbers for the four zones using first a calculator and then using the chart of Fig. 12.5. Table 12.2 gives the matrix density, bulk density, and R_t read from the logs shown. R_w may be assumed to be 0.05 Ωm.

Table 12.2 Injun Sandstone log readings.

Zone	ρ_{ma} gm/cc	ρ_B gm/cc	R_t Ohm-m	R_t/R_w	ϕ %	S_g %
A	2.75	2.36	125			
B	2.75	2.48	120			
C	2.75	2.44	21			
D	2.68	2.36	5.6			

The same log readings can also be run through the nomographic chart of Fig. 12.8, below.

Note that the red line corresponding to the value of ρ_B for zone A and the dotted line corresponding to the value of R_t/R_w. In the case shown (Zone A) the matrix density is assumed to be 2.75 g/cc and R_t is read from the resistivity log as 125 Ωm. If, as specified, R_w is 0.05, then the Y-axis on the upper right of the chart should be entered at 2500. The crossing of the dotted and red lines then leads the interpreter to the wanted values of porosity (14%) and gas saturation (87%).

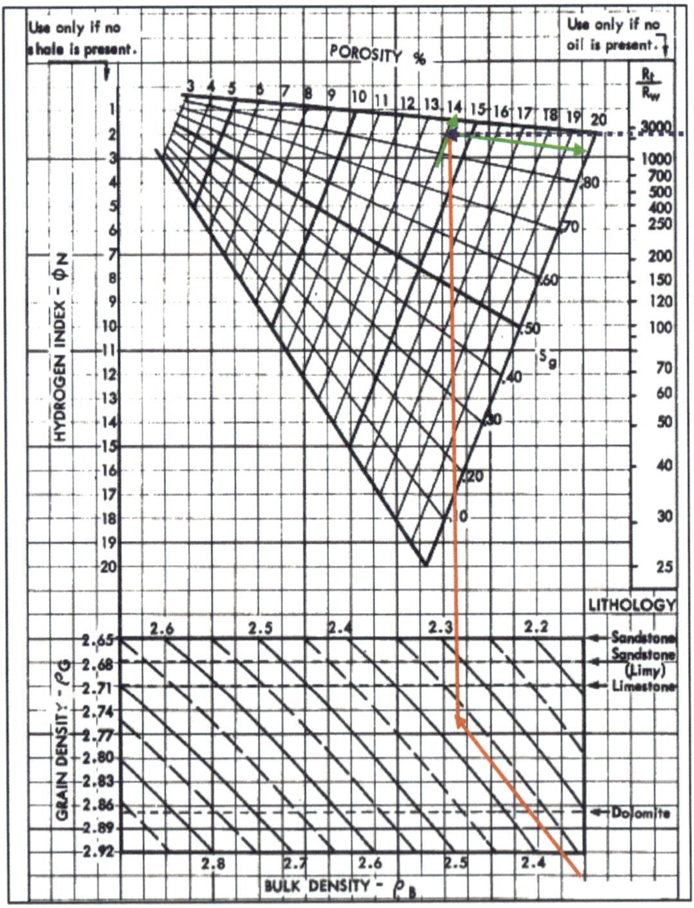

Figure 12.8 —Nomographic interpretation for empty-hole density and resistivity logs. Image courtesy Schlumberger.

Further reading

Douglas, R.D., 1947. Elements of Nomography, First Edition McGraw-Hill Book Company, Inc, New York and London.

Rodermund, C.G., Alger, R.P., Tittman, J., 1961. Empty-Hole Logging Programs for Reservoir Evaluation. Society of Petrophysicists and Well-Log Analysts (SPWLA), 2nd Annual Logging Symposium held in Dallas, Texas.

Appendix

Answers to text exercises:

Zone	ρ_B gm/c	ϕ_N %	ϕ %	S_g %
A	2.44	6.0	11.2	46.4
B	2.41	6.0	12.3	51.3
C	2.40	6.0	12.7	52.7
D	2.34	4.5	14.4	68.7

Clinton sandstone example.

Zone	ρ_{ma} gm/cc	ρ_B gm/c	R_t Ohm-m	R_t/R_w	$(R_w/R_t)^{0.5}$	ϕ %	S_w %	S_g %
A	2.75	2.36	125	2500	0.020	14.9	13	87
B	2.75	2.48	120	2400	0.020	10.6	19	81
C	2.75	2.44	21	420	0.049	13.0	37	63
D	2.68	2.36	5.6	112	0.094	15.5	61	39

Injun sandstone example.

CHAPTER 13

Quick-look methods

Abstract

Quick-look methods, by definition, allow the log analyst to find useful petrophysical formation properties without having to make any numerical calculations. This chapter summarizes several such methods as they were used in the pre-digital era. Many are based on visual comparison of analog curves on a paper log print. Some of these methods are still in use today, others have been forgotten but are worthy of being documented here because they will show up in the files of the pre-digital archives. Compatible overlays of analog curves mirror the philosophy of taking ratios to eliminate unknowns and were made use of widely in times gone by—but they still have their place today.

The Derrick floor of a standard type drilling rig—1927. *Reproduced from In a Persian Oilfield: A Study in Scientific and Industrial Development: By J. W. Williamson, London: Ernest Benn Ltd., 1927.*

13.1 Easy picks from microlog, spontaneous potential, and gamma ray

Common to all "quick-look" methods, applicable to pre-digital well logs, is the purely visual aspect that such methods offer to the analyst. For good

Formation Evaluation with Pre-Digital Well Logs
DOI: https://doi.org/10.1016/B978-0-12-820232-6.00013-3

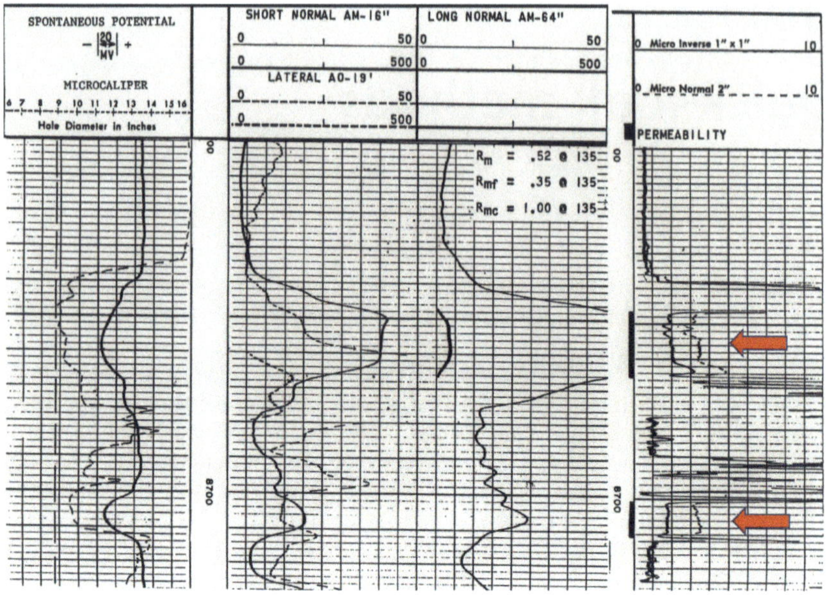

Figure 13.1 Microlog presentation indicating porous and permeable sections.

reason these methods are called "quick" and "look." A prime example of this is the Microlog. The presentation of Fig. 13.1 shows both a standard electric log along with a Microlog. Track 1, on the left, shows the micro-caliper along with a spontaneous potential (SP). Track 4, on the right shows the micro inverse and micro normal curves. In the middle is the standard E-log presentation of the short and long normals and the 18′8″ lateral. Marked with red arrows on the microlog are two zones (8647–8665 and 8700–8711), where $R_{1'' \times 1''} < R_{2''}$, that show where there is porosity and permeability. Test your skills. Which of the two zones is water bearing and which has the hydrocarbons?[1]

Equally easy to read are SP and gamma ray recordings that allow for rapid discrimination between rocks of reservoir quality and impervious zones of no further interest for the analyst. Fig. 13.2 is an example of a sand/shale sequence over a 95-ft zone.

Sands and shales are visually distinguished by the yellow shading which extends from the mid-point of the SP curve between the "sand line" and the "shale line." It serves very well as a quick-look assessment of the sedimentary sequence and allows for a painless estimate of the net-to-gross ratio for the inspected section of the wellbore.

[1] See the appendix for suggested answer.

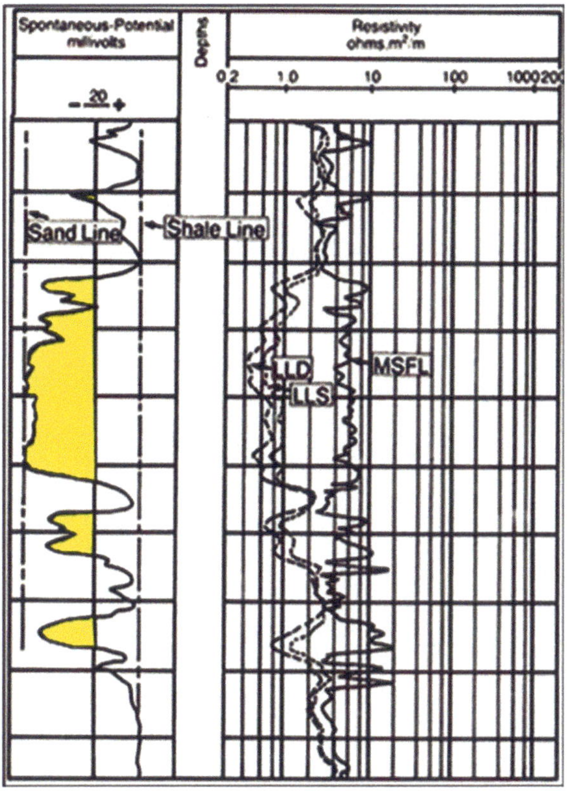

Figure 13.2 SP curve as sand/shale discriminator. *Image courtesy Schlumberger.*

Figure 13.3 Gamma ray as quick-look sand/shale gross section indicator.

Likewise, the gamma ray serves a similar purpose as illustrated in Fig. 13.3. The logged section shown covers a 1500-ft section of which approximately half is sand and half shale. If anything closer than such an "eye-ball" assessment is required, the analyst can always do some interval counting noting that the depth lines on this display are at 20-ft intervals.

13.2 R_{wa} for connate water resistivity (R_w)

The apparent water resistivity, R_{wa}, became a useful quick-look method for identifying potential hydrocarbon-bearing intervals (and the value of R_w) as soon as combination logging tools became available. With such tools it was possible to record (simultaneously) resistivity and porosity on a single trip into the hole, for example with the Induction-Sonic combination.

As a reminder we may write Archie's equation as

$$S_w^n = F\frac{R_w}{R_t}$$

which, with rearrangement of the terms, gives

$$R_w = R_t\frac{(S_w)^n}{F}$$

If S_w is now assumed to be 1, the apparent water resistivity (R_{wa}) is given by

$$R_{wa} = \frac{R_t}{F} = R_t\left(\frac{\phi^m}{a}\right)$$

The R_{wa} parameter has the charm of being exactly equal to R_w in 100% water-bearing intervals and being much larger than R_w in hydrocarbon-bearing zones, where S_w is less than 1. Fig. 13.4 gives an example of a continuous R_{wa} curve plotted on the display of a combination induction-sonic log.

The charm of this quick-look display is that the water saturation is easily calculated in any zone where the R_{wa} curve is greater than the minimum from

$$S_w = \left(\frac{R_{wa-min}}{R_{wa}}\right)^{1/2}$$

The raw measurement of the sonic tool is the transit time, Δt. In order to generate the R_{wa} number the logging equipment had to convert the raw measurement, in μseconds/ft, into porosity. This required input from the logging operator to select the constants in the Wyllie time average equation (see Chapter 4) of Δt_{ma} and Δt_{fluid}. Since the majority of induction-sonic combination logs were run in sand-shale sequences these computational constants were normally selected as 55.5 and 189 μseconds/ft, respectively. With luck the actual choices will be noted on the log heading (or footings), if they have survived to reach the modern analyst's hands.

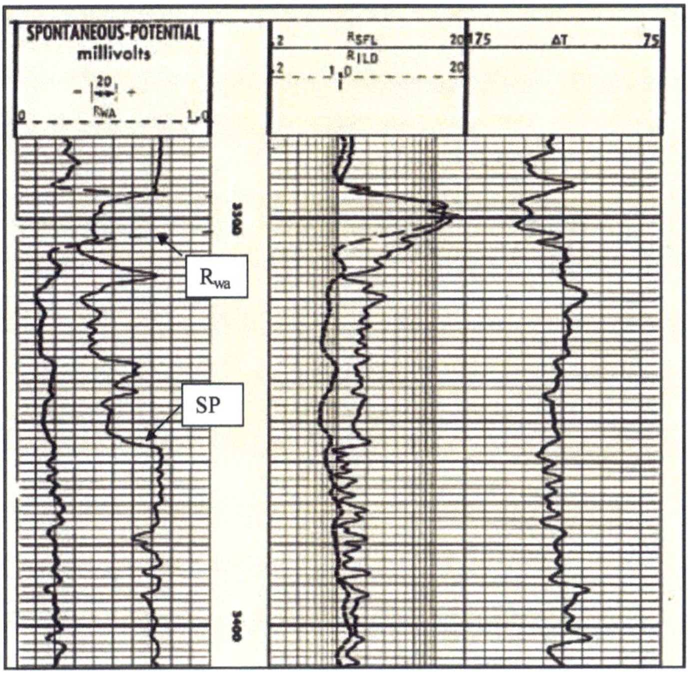

Figure 13.4 R_{wa} curve plotted on an induction-sonic combination log. Image courtesy Schlumberger.

In the example log of Fig. 13.4 the reader is directed to the zone on the log at 3300 ft depth where the dashed R_{wa} curve moves from an average of 0.05 Ωm (in the deeper section of the well) to completely off scale at 1.0 Ωm. At the 3300′ depth the sonic records a Δt of 130 μseconds/ft and the induction records approximately 12 Ωm. Sonic porosity, assuming Sandstone with a 1.3 compaction correction (shallow sediments), is 0.33. Thus, the R_{wa} value is actually $12 \times (0.33)^2 = 1.3$ Ωm—although we cannot see it, since it is off scale. The minimum value of the R_{wa} curve in what appears to be the clean water-bearing section (at 3326′) is 0.05 Ωm. A quick calculation thus gives an apparent water saturation in the "pay" zone of $(0.05/1.3)^{1/2} = 0.195$ or 19.5%.

13.3 Continuous R_{xo}/R_t versus SP analog curve plot

Chapter 10 describes in detail the "Rocky Mountain Method" which is a perfect example of the ingenuity of the early log analysts. When faced with few "knowns" and many "unknowns" they resorted to clever devices to eliminate entirely what was not known by taking ratios. What was left was a useful answer product that solved many log analysis problems. Tixier published his paper in 1949 which showed that it was possible to deduce water saturation with the use of a resistivity/SP log alone, without the need for a porosity log. For those who have skipped Chapter 10 the derivation of the formula for S_w is repeated here. It will help to underscore the simplicity and power of the method. We start by writing down Archie's equation for both the invaded and uninvaded zones:

$$S_w{}^2 = \frac{FR_w}{R_t} \text{ and } S_{xo}{}^2 = \frac{FR_{mf}}{R_{xo}}$$

Since porosity, and hence the formation factor F, are unknowns, a ratio of these two equations eliminates the unknown quantity as follows:

$$\frac{S_w{}^2}{S_{xo}{}^2} = \frac{R_w}{R_{mf}} x \frac{R_{xo}}{R_t}$$

An assumption may be made about the relationship between invaded and uninvaded water saturations and, based on core flooding experiments and other observations, a general equation sets S_{xo} equal to $S_w{}^{(1/5)}$. The left-hand side of the equation now simplifies to

$$S_w{}^{(8/5)} = \frac{R_w}{R_{mf}} x \frac{R_{xo}}{R_t}$$

The final piece of the jigsaw puzzle is provided by the SP, which also depends on the ratio of the connate water resistivity, R_w, to the invading mud filtrate resistivity R_{mf}.

$$\text{SP} = -K \log \frac{R_{mf}}{R_w}$$

From which we derive as

$$\frac{R_w}{R_{mf}} = 10^{(\text{SP}/K)}$$

which leads to the final expression for S_w as

$$S_w = \left[10^{(\text{SP}/K)} x \frac{R_{xo}}{R_t}\right]^{5/8}$$

As was shown in Chapter 10 this equation can be solved graphically on an $X-Y$ plot with the (R_{xo}/R_t) ratio plotted against the SP. An even more elegant method of making a comparison between the two parameters is to plot them continuously on a log. A properly scaled version of the R_{xo}/R_t ratio will then track the SP curve when the formation is 100% water filled and will depart from the SP when the formation is hydrocarbon bearing. Fig. 13.5 illustrates the method.

The dual induction tool was the first available to produce this kind if a display since it could record both R_t and R_{xo}, (or a close approximations thereto) and also had surface equipment sufficiently advanced to continuously take a ratio between the two and plot the required $(R_{xo}/R_t)_{QL}$ along with the SP in real time. In the example of Fig. 13.5 the tool used was a dual induction-LL8. The ILd measurement was taken as a measure of R_t and the LL8 as a measure of R_{xo}.

These kinds of plots are to be found in many old log files. They may not be as pretty as this one and the trade craft at the wellsite was not always perfect so, on occasions, the SP and the $(R_{xo}/R_t)_{QL}$ did not always

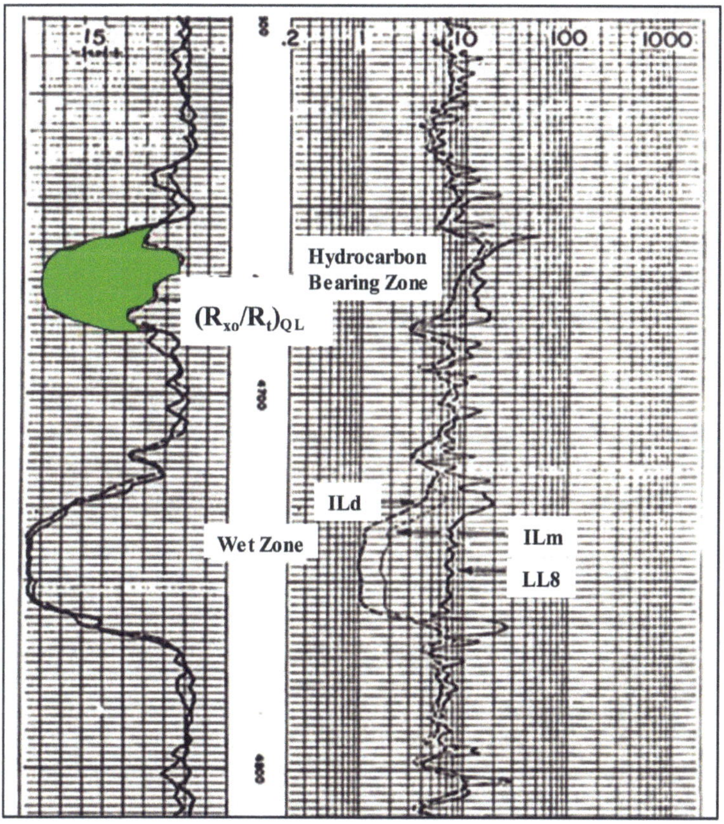

Figure 13.5 (R_{xo}/R_t) "quick-look" and SP curve presentations for $S_w < 100\%$. *Image courtesy Schlumberger.*

line up exactly in the water-bearing zone. However, a modern analyst can always make adjustments with a little tracing paper and achieve the required result—a quick-look assessment of possible hydrocarbons using just a resistivity log and without the help of either a porosity log or any prior knowledge of R_w or R_{mf}. Truly a neat trick.

This method has applications where there is a good contrast between the salinity of the formation water and the salinity of the mud filtrate. It is this contrast that drives the SP in the first place and where the two salinities are close the SP deflection from a shale line to it maximum in a porous bed will be small and the method will lose resolution and visual impact.

13.4 F-overlay for water saturation

The F-overlay was another popular wellsite curve overlay technique that was widely used in the late 1960s and early 1970s. It was based on the simple Archie principal that water saturation can be deduced from a knowledge of R_t, the actual rock resistivity, and R_o, the resistivity that same rock would have if it were 100% saturated with water:

$$S_w = \sqrt{\frac{R_o}{R_t}}$$

Getting R_t is straight forward. A deep resistivity device provides this, either from a deep induction in fresh mud systems or from a laterolog in salt mud systems. Getting a continuous curve of R_o is a little more complicated. The way it was done was to start with a porosity curve and massage it into a formation factor (F) curve. Typically this would be done with a neutron (mostly at that time sidewall neutron porosity [SNP]) or a bulk density (ρ_B) curve and using the surface equipment in the logging truck to make the necessary conversions from raw porosity measurement to F:

$$F = \frac{a}{\phi^m}$$

The final logical step is to change F into R_o and this is accomplished by multiplying F by R_w:

$$R_o = FR_w$$

Fig. 13.6 shows a section of log with both R_o and R_t curves overlaid to coincide in a water-bearing zone and separated in the hydrocarbon-bearing zone. The "normalization" of the F curve to become an R_o curve is the logarithmic equivalent of a multiplication. The log of F plus the log

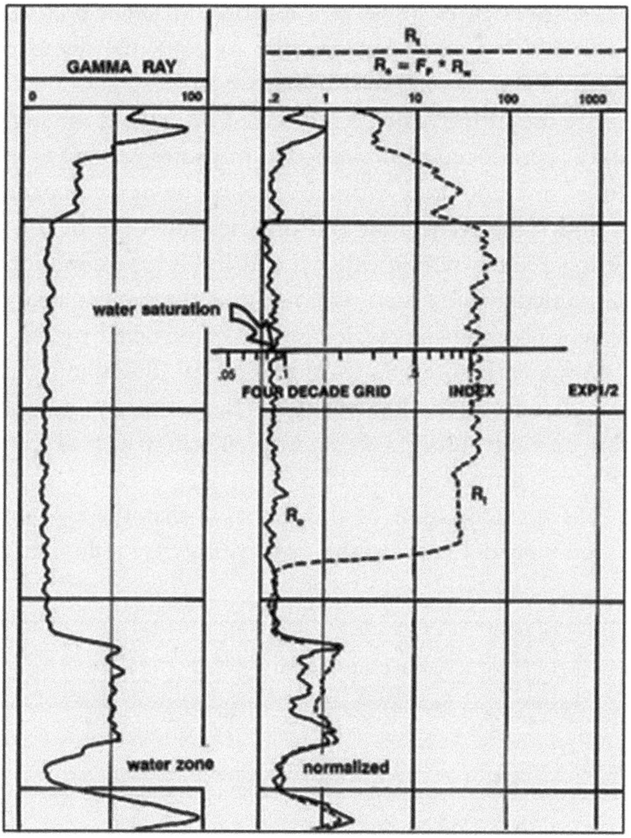

Figure 13.6 R_o versus R_t (F-overlay). *Image courtesy Schlumberger.*

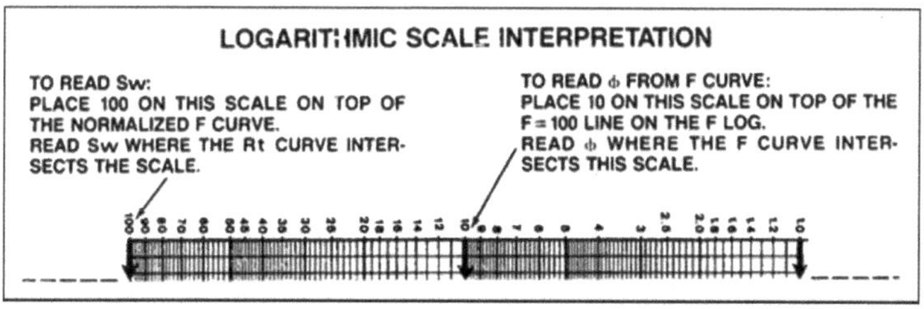

Figure 13.7 Logarithmic scaler for reading S_w from R_o/R_t overlay *(not to scale).*

of R_w will give the log of FR_w, or log R_o. The distance on the plot between the R_o and the R_t curves in the hydrocarbon-bearing zone is directly related to the ratio of R_o to R_t. Remembering that the water saturation is given by the square root of the ratio of R_o to R_t, a special scaler can be used to read the water saturation directly from the overlay plot itself. Fig. 13.7 gives an example of a logarithmic scaler that can be used on a full-sized API log grid print. For this to work the distance between the 100 and the 1.0 marks has to be exactly 5 in.

The correct use of the scaler is predicated on having the log print and the scaler itself reproduced to the exacting measures needed to ensure that the "slide rule" methodology works. Wireline service companies offered such scalers and if the reader can find one, so much the better. However if the paper log print is not an original and has been reproduced with an optical scale reduction the system will not give the correct answers for S_w. In such a case a "homemade" scaler can be constructed by following the procedure shown in Figs. 13.8, 13.9, and 13.10. The starting point is to mark off two decades of the log scaling for the resistivity log under analysis. This can be done using a 6-in. strip of stiff paper or cardboard as shown by the two arrows on Fig. 13.8.

Step 2 calls for a rotation of the scaler so that the two marks now encompass just *one* decade on the base resistivity scale. As shown in

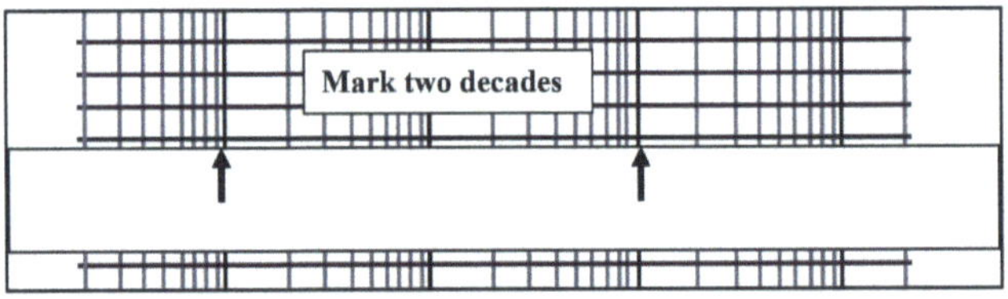

Figure 13.8 Step 1 for procedure for custom made F-log scaler.

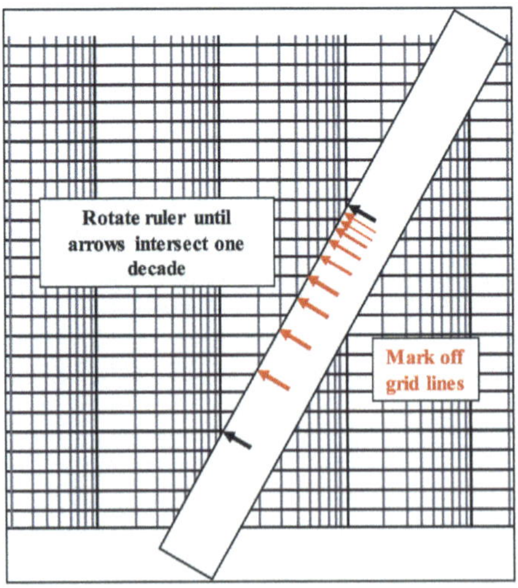

Figure 13.9 Step 2 for procedure for custom made F-log scaler.

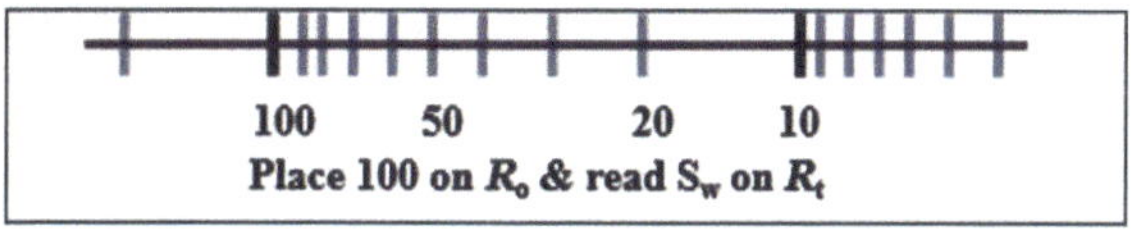

Figure 13.10 Step 3 for procedure for custom made F-log scaler.

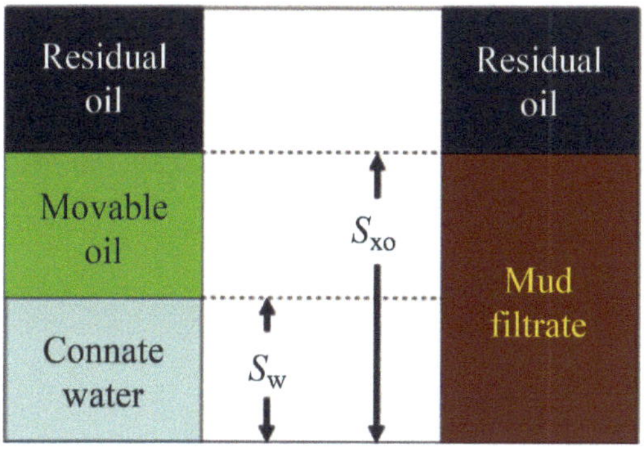

Figure 13.11 Basis for the movable oil plot.

Fig. 13.9 the log scale lines from 1 to 10 (or 10 to 100) can now be ticked off on the 6-in. strip.

Step 3 calls for the scaler to be flipped and marked off as shown in Fig. 13.10. Thereafter the water saturation at any depth on the log can be found by lining up the 100 on the scaler with the R_o curve and reading where the R_t curve crosses the scaler.

13.5 Movable oil plot

In those cases where a logarithmically scaled R_{xo} curve is available it is also possible to expand the F-overlay method to produce what is referred to as a movable oil plot (MOP). The basis for the method is a comparison between the water saturation in the undisturbed zone (S_w) and the water saturation in the invaded zone (S_{xo}). As shown in Fig. 13.11 the difference between the two saturations corresponds to the movable oil present.

To prepare an MOP using the F-overlay technique requires a logarithmic R_{xo} recording. If original film recordings are unavailable, then paper prints can be used with a light table to trace the normalized R_o and R_{xoo}[2] curves on to the main R_t display. Just as R_t is "normalized" in the 100%

[2] R_{xoo} is the value of R_{xo} when $S_{xo} = 100\%$.

water-bearing zone by an amount equal to the log of R_w, the R_{xo} is "normalized" in the same zone by an amount equal to the Log of R_{mf}. Note that the normalization is effectively a slide rule version of taking the square root R_o divided by R_t (the subtraction of two logarithms). The same applies to the S_{xo} determination by taking the square root of R_{xoo} divided by R_{xo}. A complete logarithmic MOP is illustrated in Fig. 13.12.

The MOP was a convenient way of giving a visual display of the distribution of hydrocarbons and water in the pore space without the use of digital log data. At the end of the pre-digital era, when digital logs data and computer processed interpretations (CPIs) become the standard, the

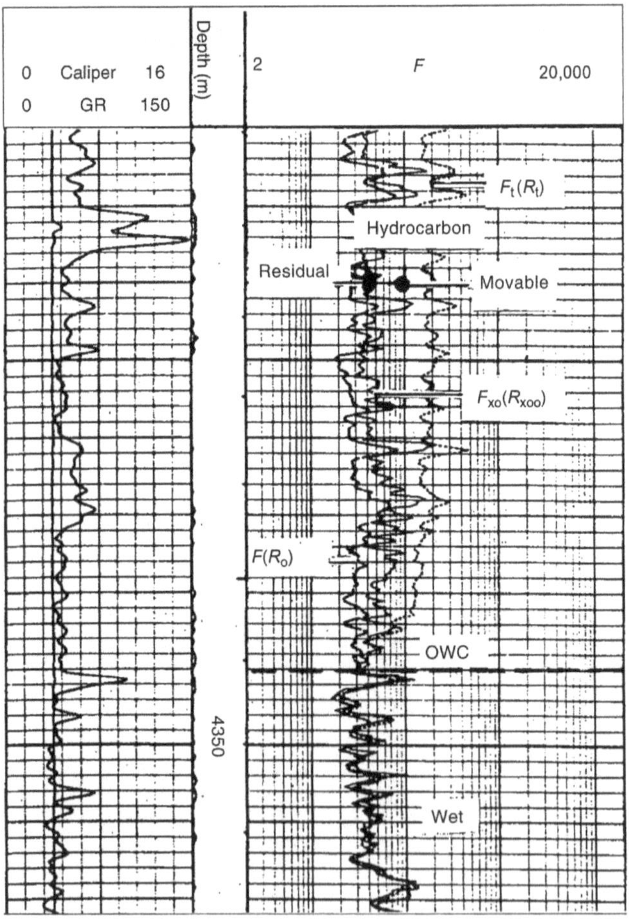

Figure 13.12 Logarithmic movable oil plot. *Image courtesy Schlumberger.*

same presentation of ϕ, ϕS_w, and ϕS_{xo} was used and still is today. Chapter 18 covers the making and presentation formats of CPIs in more detail.

13.6 Density/sonic and resistivity/sonic comparisons for fractures

The basis for the density/sonic quick-look overlay for fracture detection lies in the basics of measurement of the two devices. The density tool responds to the total porosity and is unable to distinguish between matrix (intergranular) porosity and fracture or vugular porosity. Fig. 13.13 illustrates this principle.

The sonic travel time measurement is based on the first arrival of a compression wave traveling from the sonic transmitter through the formation following the fastest route. Since travel time is faster through the matrix and slower through a fluid-filled fracture, the sonic "porosity" reflects the matrix contribution and ignores any secondary porosity. Thus the apparent porosities from density and sonic will differ where there are fractures. In Chapter 4, which covers Sonic logging and interpretation more fully, the compatible "quick-look" overlay of sonic travel time, Δt, and bulk density, ρ_B, is reproduced here as Fig. 13.14. This shows how the combination of raw log measurements can highlight any secondary porosity. Density porosity will appear higher than sonic porosity in

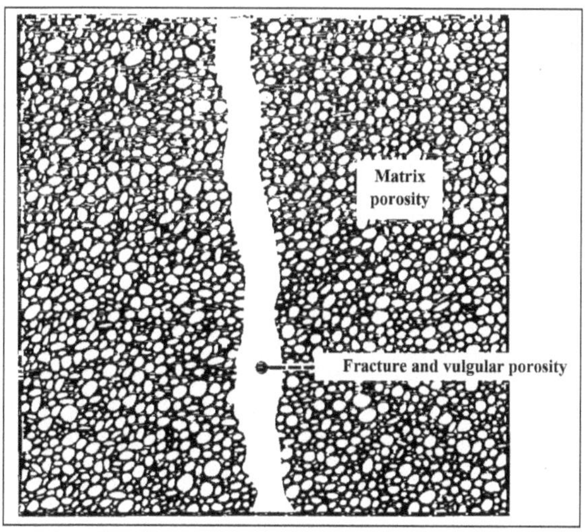

Figure 13.13 Matrix and vugular porosity.

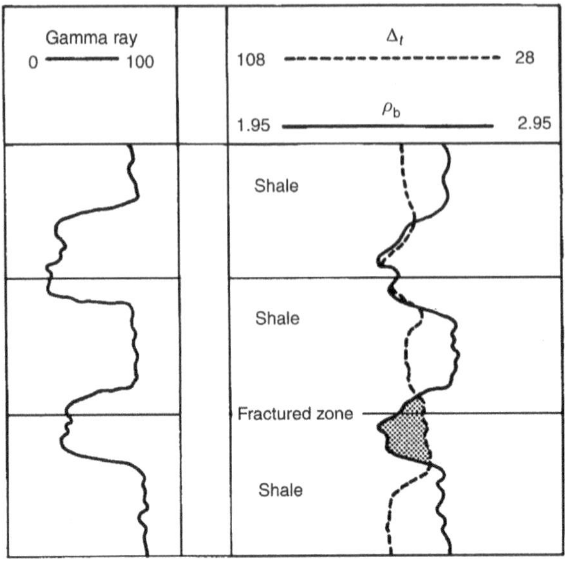

Figure 13.14 Δt and ρ_B overlay for fracture detection quick look.

fractured formations. This same principle is used in CPIs as discussed in Chapter 18. Secondary porosity in those applications is defined as

$$\phi_{\text{Secondary}} = \phi_{\text{Density}} - \phi_{\text{Sonic}}$$

Another quick-look fracture detection technique that may be found in well log archives relies on a compatibly scaled sonic amplitude measurement (effectively a cement bond log run in open hole) together with a formation resistivity measurement, such as an amplified 16″ short normal. This presentation was known as a "Many Special," so named after the field in Louisiana where it was first used.

A further simple and direct method of inferring the presence of vugs and fractures can be obtained from an analog print of a four-arm dipmeter log. Fig. 13.15 gives an example of this method.

Two independent calipers are free to record two hole diameters at right angles to each other. In general, boreholes are not circular but oval and the long axis is usually associated with any breakout due to fracturing. Thus the actual resistivity values recorded by the electrodes on each of the four caliper arm pads will reflect either high resistivity in front of solid rock or a slightly lower resistivity in front to a fluid-filled fracture or other vugular fissure. Thus, by superimposing the resistivities from adjacent pads (e.g., 1 cf. 2 and 3 cf. 4), differences will show up in fractures. Note on

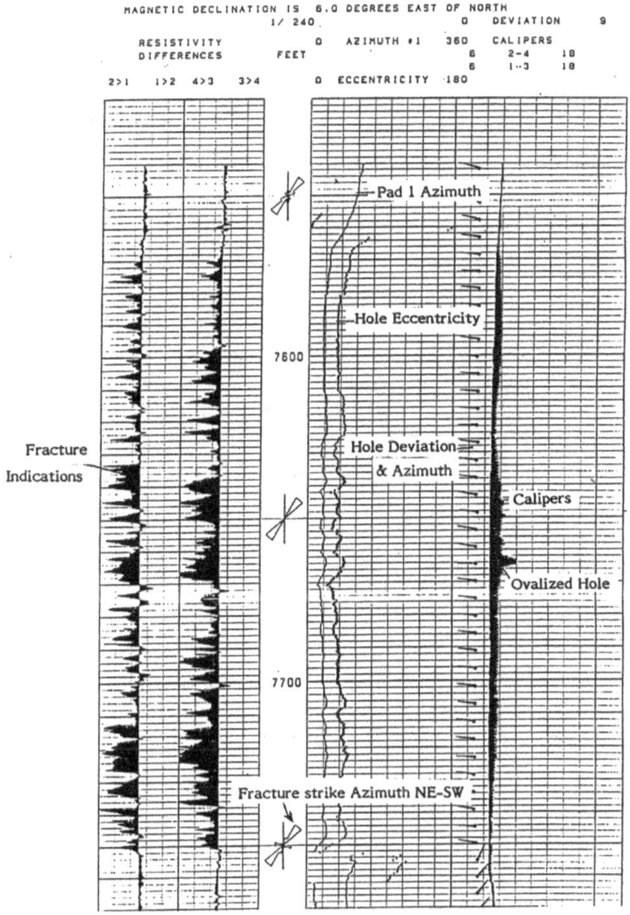

Figure 13.15 Dual four-arm dipmeter calipers used for fracture identification.

the example of Fig. 13.15 that the main fractures zone is between 7500 and 7850 ft. The two calipers also indicate hole ovalization between 7500 and 7655.

13.7 Neutron/density quick looks for gas, oil, and lithology

Once combination tools were introduced into the wireline logger's arsenal it became common to run compatible density-neutron displays to highlight either hydrocarbons or lithology.

In sand/shale sequences these displays were scaled in sandstone porosity units from zero porosity to 60% across two tracks, as shown in

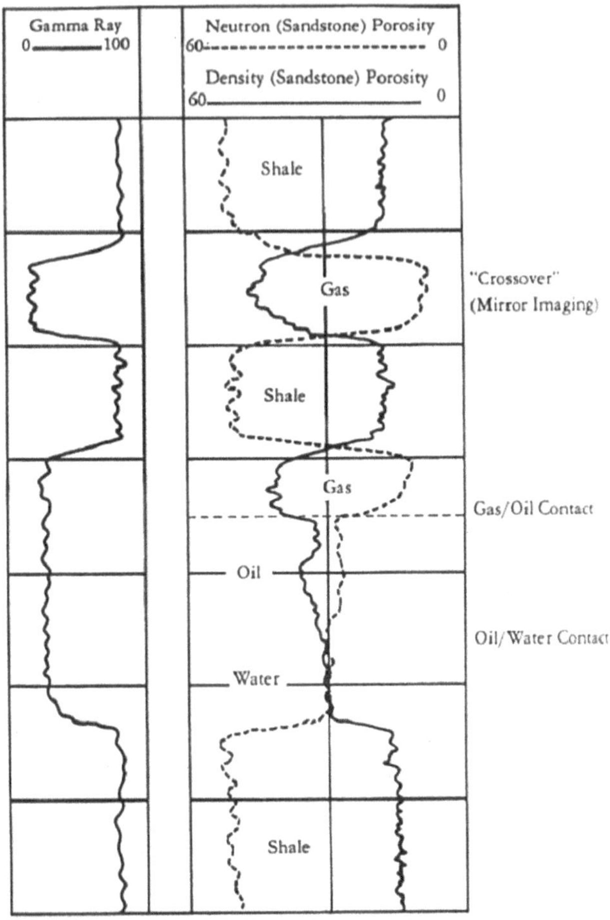

Figure 13.16 Neutron density overlay (sandstone scaling) for gas and oil quick look.

Fig. 13.16. "Crossover," with neutron greater than density, indicates shale, whereas neutron less than density indicates oil or gas. In clean water-bearing sandstone both read the same.

Fig. 13.17 illustrates the sister plot more commonly found in "hard rock" territory. Here both the neutron and density are displayed on apparent limestone porosity scaled from −15% to 45% (right to left) across two tracks. Note, on Fig. 13.17, that neutron is greater than density in shale, anhydrite and dolomite but less in salt and sandstone. Only in limestone will the two curves coincide. The scaling of −15 to +45% is dictated by the density of anhydrite (2.98 g/cc) that has an apparent limestone porosity of very close to −15% as given by

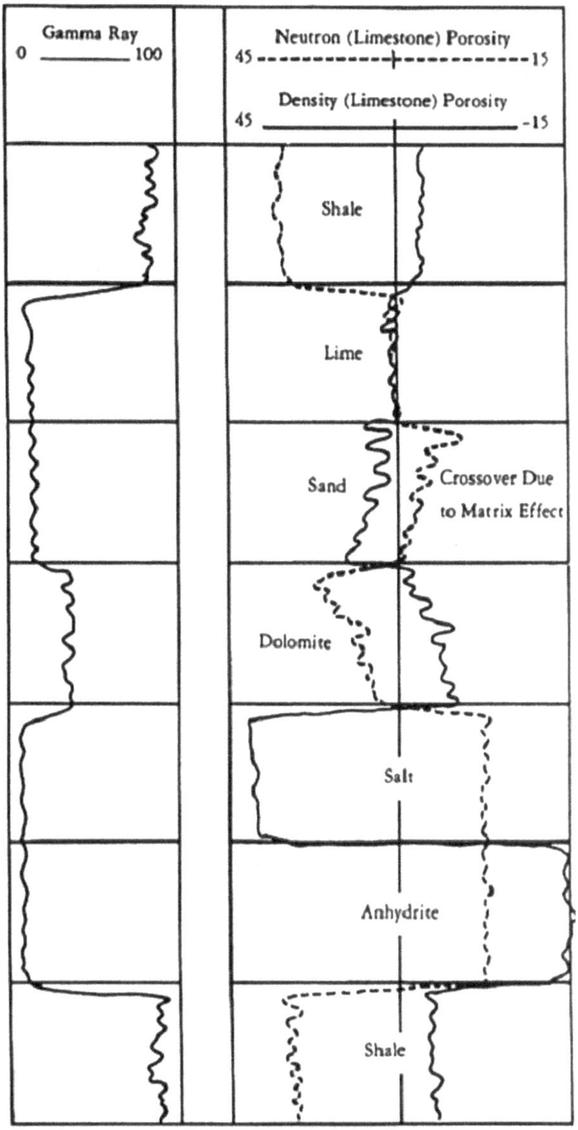

Figure 13.17 Neutron density overlay (limestone scaling) for lithology quick look.

$$\phi_D = \frac{2.71 - 2.98}{2.71 - 1.00} = \frac{-0.27}{1.71} = -0.158$$

Quick-look methods certainly have an important role in formation evaluation from pre-digital well logs. Their value lies in the time and effort saved by the avoidance of costly analog curve digitization and subjection to "modern" CPI. Use of the quick-look methods can show that CPI may not be merited if a first scan using any or all of the techniques described in this chapter eliminates unpromising zones from further consideration. But they may equally highlight the most promising zones that may merit more detailed analysis.

Further reading

Buckles, R.S., 1965. Correlating and averaging connate water saturation data. J. Can. Petrol. Technol. 9 (1), 42–52.

Gael, T.B., 1981. Estimation of Petrophysical Parameters by Crossplot Analysis of Well Log Data. Louisiana State University, Baton Rouge, Louisiana, MS Thesis.

Krygowski, D.A., Cluff, R.M., 2015. Pattern recognition in a digital age: a gameboard approach to determining petrophysical parameters. SPWLA 56th Annual Logging Symposium held in Long Beach. 18–22 July, California, USA.

Morris, R.L., Biggs W.P., 1967. Using log-derived values of water saturation and porosity. Society of Professional Well Log Analysts, 8th Annual Symposium Transactions, Paper X.

Pickett, G.R., 1966. Review of current techniques for determination of water saturation from logs. J. Petrol. Technol. Soc. Petrol. Eng. 18 (11), 1425–1433.

Pickett, G.R., 1973. Pattern recognition as a means of formation evaluation. SPWLA Fourteenth Annual Logging Symposium, Paper A.

Tixier, M.P., Alger, R.P., Tanguy, D.R., 1960. New developments in induction and sonic logging, Society of Petroleum Engineers. J. Petrol. Technol. 12 (5), 79.

Tixier, M.P., Eaton, F.M., Tanguy, D.R., Biggs, W.P., 1965. Automatic Log Computation at Wellsite-Formation Analysis Logs. Society of Petroleum Engineers. 17 (1). Available from: https://doi.org/10.2118/987-PA.

Schlumberger Ltd, 1971. The ISF/Sonic, Technical Report. Schlumberger Ltd.

Appendix

Answer to question posed regarding Fig. 13.1:

Zone at 8647–8665 hydrocarbon bearing.

Zone at 8700–8711 water bearing.

SECTION V

Permeability estimates

CHAPTER 14

Permeability estimates

Abstract

Today there are many high-tech methods for measuring formation permeability, either by promoting fluid movement and gauging draw-down pressures or by flipping protons with a nuclear magnetic resonance tool. However, there still exist a number of proven, reliable methods for calculating permeability even from the oldest of the old, pre-digital logs. The summary of these analog methods, given here, will serve to remind today's analyst that the way we got to see so far today is because we stand on the shoulders of giants[1].

Early logging operations in South America. Image courtesy Schlumberger

14.1 Classical methods for permeability estimating

Today we are fortunate to have access to tools and techniques that were not available to the pioneers of formation evaluation. They had no nuclear magnetic resonance (NMR) log and no wireline formation testers with pretest capabilities. What they did have, instead, was reliance on basic petrophysical relationships at the pore scale and their connections to log measurable entities such as irreducible water saturation. This, coupled with grunt work with

[1] With due thanks to Isaac Newton, of course.

Formation Evaluation with Pre-Digital Well Logs
DOI: https://doi.org/10.1016/B978-0-12-820232-6.00014-5

core samples, allowed them to make thoughtful correlations of basic logging measurements to the elusive estimation of permeability.

In this section, we will revisit some powerful but sometimes forgotten methods of permeability estimating. They still have a place today in the arsenal of the petrophysicist for insurance against the day when, for whatever reason, the modern methods are not available or affordable.

14.2 Resistivity gradients

One of the earliest methods relied on a simple resistivity log alone. As more fully explained elsewhere,[2] the column of fluids in the transition zone between an oil–water contact and the point where irreducible water saturation is reached shows a characteristic function of water saturation against depth and an equally characteristic shape to a conventional formation resistivity curve. Fig. 14.1 shows these curve shapes. Note, on the diagram, that the water saturation (blue line) changes from 100% at the oil–water contact to its irreducible value ($S_{w\text{-}irr}$) at the top of the transition zone. In the case illustrated, $S_{w\text{-}irr}$ is 32%. Between the two limits, the saturation follows a curved path. Over the same interval, the formation resistivity (green line) follows a straight line on this plot of depth in feet versus R_t in Ωm. The gradient is quantifiable in ohm meter per foot. High-permeability formations produce short transition zones, which are characterized by low-resistivity gradients. Low-permeability formations

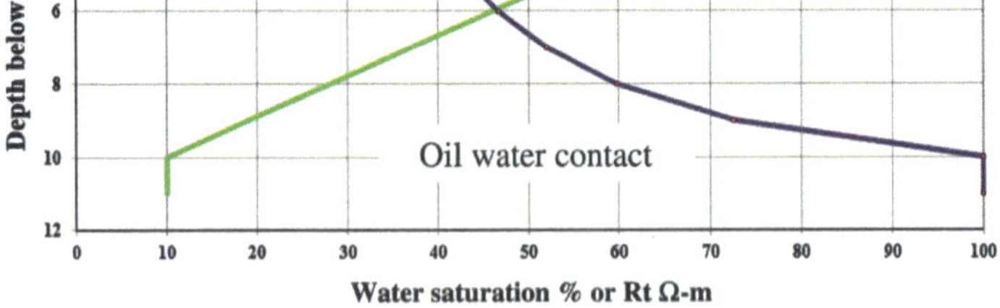

Figure 14.1 Water saturation and formation resistivity in a transition zone.

[2] See Bateman's Openhole Log Analysis and Formation Evaluation, Chapter 6.

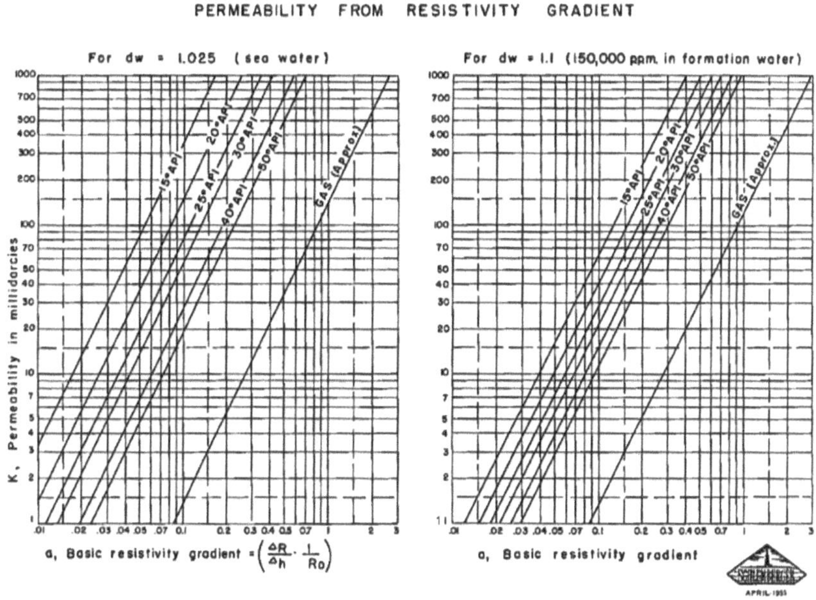

Figure 14.2 Permeability from resistivity gradient. *Image courtesy Schlumberger.*

produce long transition zones with high-resistivity gradients. The key to quantitatively gauging formation permeability is thus to link the slope of the resistivity ($\Delta R/\Delta h$) to permeability.

An early publication achieved this by use of two charts that are reproduced here as Fig. 14.2. Note that the depth difference between the OWC and the top of the transition zone depends not only on the permeability of the formation but also on the fluids present and their densities. This is just a reflection of the basic equation that states:

$$h = \frac{2T\text{Cos}\theta}{r\left(\rho_{\text{w}} - \rho_{\text{hy}}\right)}$$

where h is the height above the OWC, T is the surface tension, θ is the contact angle, r is the radius of the capillary tube, ρ_{w} is the water density, and ρ_{hy} the hydrocarbon density. Thus, a given resistivity gradient is dependent on both the formation permeability and the fluids in the reservoir. Inspection of Fig. 14.2 will reveal that it is built in two panels. The left-hand side is built for sea water salinity reservoirs ($\rho_{\text{w}} = 1.025$ g/cc) and the panel on the right for saltwater reservoirs ($\rho_{\text{w}} = 1.1$ g/cc). On both the analyst enters with the "Basic Resistivity Gradient," which is defined as:

$$a = \frac{\Delta R}{\Delta D} x \frac{1}{R_{\text{o}}}$$

where R_o is the resistivity of the 100% water bearing formation below the OWC. A line is then extended up the chart to intersect the API gravity of the oil in the reservoir (or to the gas line), and the permeability is then read in milli-Darcy (mD) on the *Y*-axis. The full equation solved graphically by these charts is:

$$k = 20\left\{\frac{2.3a}{\left(\rho_{\mathrm{w}} - \rho_{\mathrm{hy}}\right)}\right\}^2$$

k is in mD, fluid densities in g/cc, and the resistivity gradient is in Ωm/ft.

Fig. 14.3 illustrates the method with reference to a sandstone reservoir at a depth of 6790′. In this example, it is assumed that the 64″ normal is a

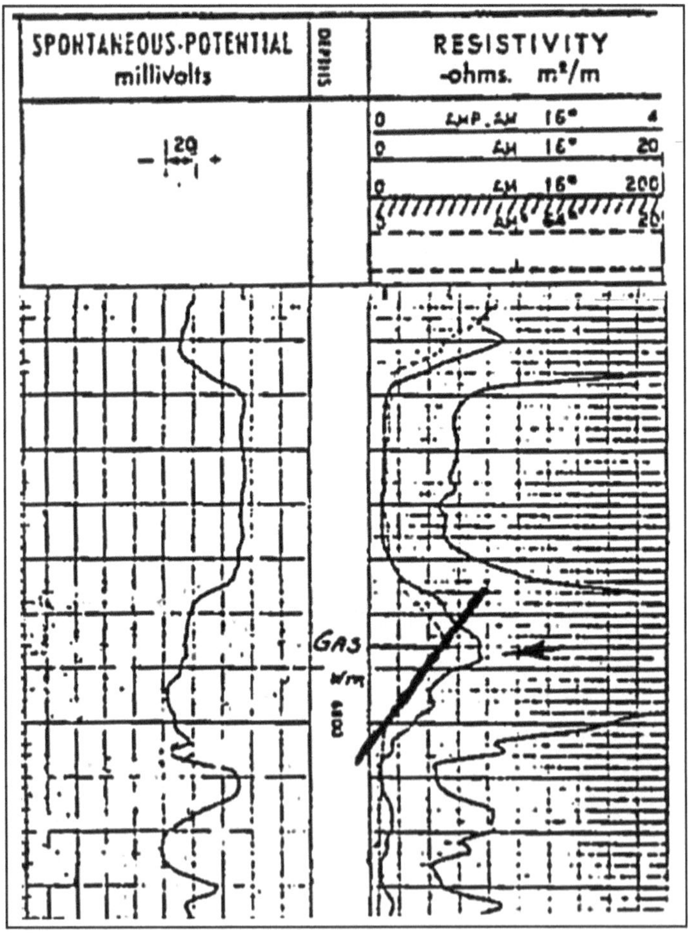

Figure 14.3 Example log for *k* from resistivity gradient.

good estimate of R_t. Then a quick analysis of the section between 6786′ and the 100% water point at 6802′ allows a calculation of the resistivity gradient as follows:

R_t @ 6786 = 5.0 Ωm

R_o @ 6802 = 0.8 Ωm

$\Delta R = 5.0 - 0.8 = 4.2\ \Omega m$

$\Delta D = 6802 - 6786 = 16'$

$$a = \frac{\Delta R}{\Delta D} x \frac{1}{R_o} = 0.33$$

Then, assuming $\rho_w = 1.1$ (right-hand chart) and the hydrocarbon presents is gas $k = 15$ mD.

14.3 Alternative graphical solution for *k* from resistivity gradient

An alternative graphical solution for permeability from the resistivity gradient equation is offered by the chart of Fig. 14.4. Here the resistivity

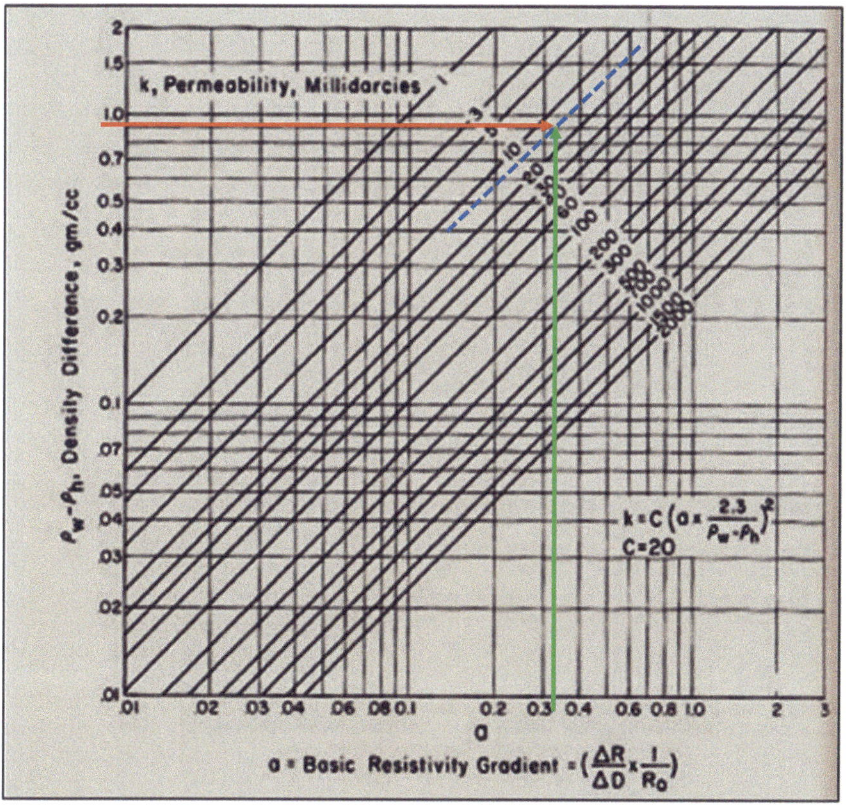

Figure 14.4 Permeability from resistivity gradient. *Image courtesy Schlumberger.*

gradient remains on the X-axis, but the Y-axis is scaled in density difference (in g/cc) between the water and the hydrocarbon phases present. Lines of constant permeability are drawn diagonally across the chart. Where plotted values intersect, the permeability is read. Low-perm lines are to the upper left and high-perm ones to the lower right. The example taken from the log of Fig. 14.3 requires the same X-axis input of 0.33 but a Y-axis input of 0.9 g/cc, corresponding to a density difference of 1.1−0.2. The plotted point is shown on Fig. 14.4. The same permeability value of 15 mD results.

14.4 The Timur equation

After an exhaustive study, based on the core analysis, of many fields and different productive formations, Timur published his results in the form of an equation that came to be known as *The Timur Equation*. It relates permeability to porosity and irreducible water saturation and is stated as follows:

$$k = 0.136\frac{\phi^{4.4}}{S_{wirr}^{2}}$$

It should be noted that both porosity and water saturation are in % in this version of the equation and the permeability is in mD. Fig. 14.5 shows a graphical representation of the Timur equation.

14.5 Wyllie and Rose

Similar in form to Timur's equation are the two equations published by Wyllie and Rose. They are as follows:

$$k = \left(\frac{250\phi^{3}}{S_{wi}}\right)^{2} \text{(Oils)}$$

and

$$k = \left(\frac{79\phi^{3}}{S_{wi}}\right)^{2} \text{(Dry gas)}$$

Note that in the Wyllie and Rose equations, both the porosity and water saturation are in fractional units. A modified version of these

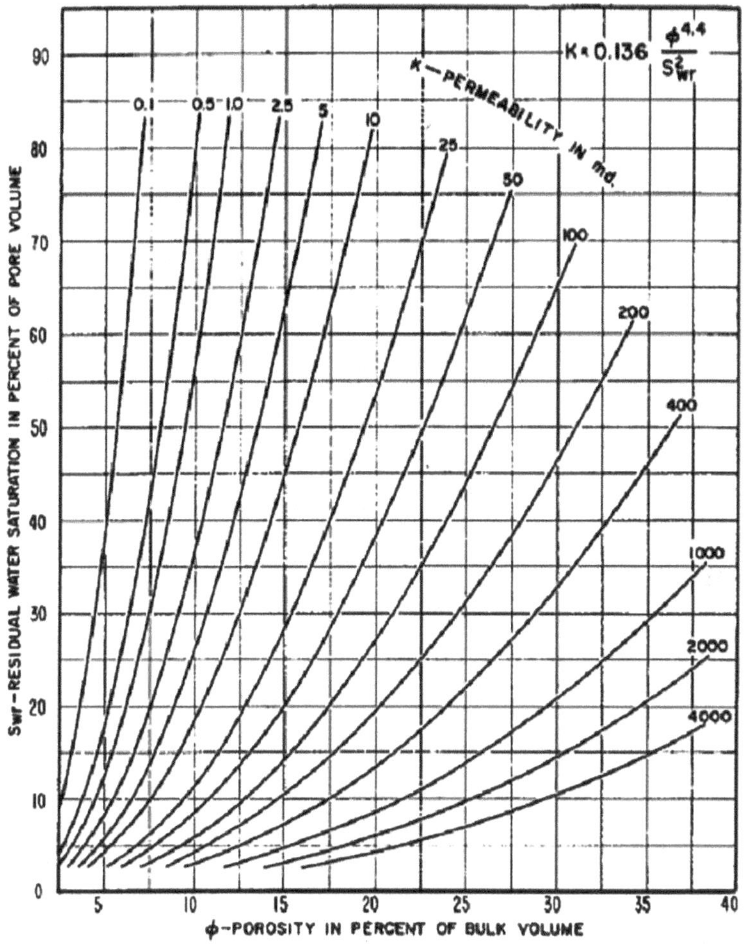

Figure 14.5 Graphical representation of the Timur equation.

equations was included in the Schlumberger Chartbooks as shown in Fig. 14.6.

Noticeable on the plot of Fig. 14.6 are the secondary lines of constant $\phi S_{\text{w-irr}}$ product. These are significant additions to the petrophysicist's arsenal. When a formation is above the transition zone, and thus at irreducible water saturation, the product of ϕ and S_w is a constant referred to as the *bulk volume water at irreducible* (BVW). Variations in porosity are normal on a local scale. If reduced, a greater proportion of the pore throats are small ones or there are simply fewer pore throats. Either way, the mean radius *r*

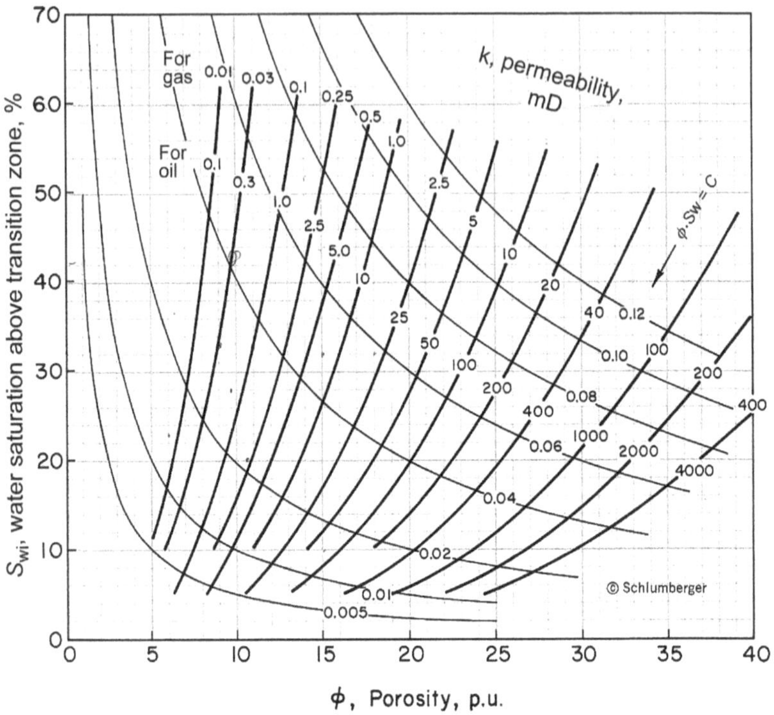

Figure 14.6 Permeability from porosity and irreducible water saturation – *after Wyllie and Rose* (Image courtesy Schlumberger)

is smaller, and more water can be held in the pore maintaining the constant ϕ $S_{w\text{-}irr}$ product.

Reservoirs may thus be characterized by the *BVW* product, and this can be used as a basis for predicting production characteristics. For points not at irreducible saturation, some water production is to be expected depending on the mobility ratio ($k_w\mu_o/k_o\mu_w$) for the particular fluids present. Fig. 14.7 shows the ϕS_w product at irreducible saturation for a number of formations. Note that in a low-porosity, low-permeability formations, surprisingly high water saturations can be tolerated without fear of water production. Conversely, in formations with good porosity and permeability, even when they have moderate values of S_w, water production can be expected.

An interesting variation on this theme is the empirical permeability chart that was published by Elgen. Fig. 14.8 shows $S_{w\text{-}irr}$ on the *Y*-axis and lines of constant ϕ across the body of the plot and the output read as *k* on the *X*-axis.

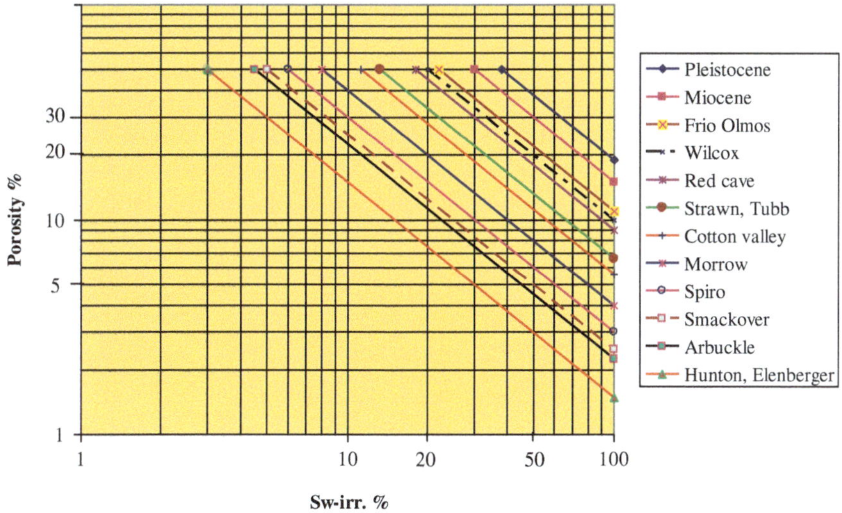

Figure 14.7 Lines of constant ϕS_w at irreducible for common formations.

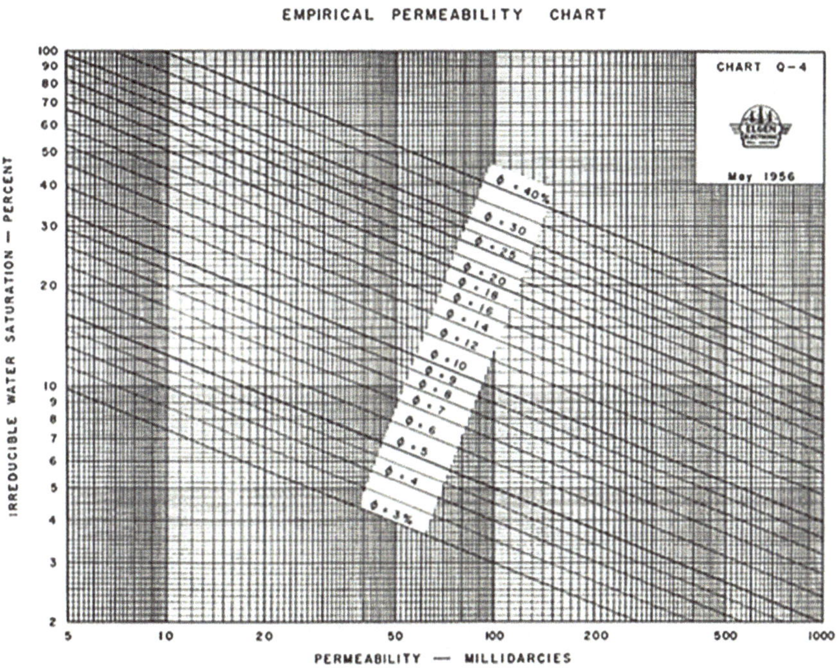

Figure 14.8 Alternative permeability chart. *After Elgen.*

14.6 Kozeny–Carman correlation

The Kozeny–Carman correlation relates permeability to both porosity and specific surface area of a porous system. Hopefully porosity can be derived from pre-digital logs as covered in Chapters 4 through 7. However, the specific surface area is not something that is simple to measure or to calculate from any known well logs. All is not lost; however, since the basic Kozeny correlation is a combination of a porosity-dependent function and a pore structure constant. The first is simple to compute from:

$$k = \text{Constant} \times \frac{\phi^3}{(1-\phi)^2}$$

Fig. 14.9 plots the basic Kozeny function with an arbitrary constant value set to 10^5.

The controlling parameter in any rock system is the pore throat radius. That is what fundamentally controls permeability and hence fluid flow. A good indicator of pore throat radius, in turn, is the irreducible water saturation. If $S_{\text{w-irr}}$ is high, then the pore throats are small and vice versa. High-perm reservoirs have low irreducible water saturations. This opens the door for use to be made of pre-digital porosity estimates to be coupled with water saturation estimates above the top of the transition zone to set up a local calibration for a Kozeny-type permeability algorithm. The shape of the porosity versus permeability graph (Fig. 14.9) will be the same; just the values on the *Y*-axis will change depending on local calibration. This is effectively just a question of moving the whole curve, shown on Figure 4.9, up or down the page depending on the characteristics of the flow unit of interest.

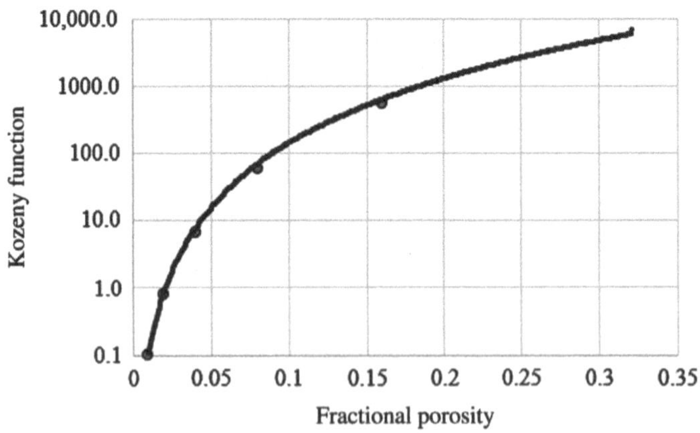

Figure 14.9 Basic Kozeny function versus porosity.

14.7 Raymer and Freeman

A variation on the resistivity gradient method for determining permeability is offered by the Raymer and Freeman method. They used a general expression based on the porosity and transition zone length, that is, the height in the reservoir from the free water level to the top of the transition zone where irreducible water saturation is reached. They offered two versions of their formula, one for oil–water cases and another for gas–water. Their equations are as follows:

$$k = \phi\left\{\frac{122}{h(\rho_{\mathrm{w}} - \rho_{\mathrm{o}})}\right\}^2 \text{ for water-oil systems}$$

and

$$k = \phi\left\{\frac{140}{h(\rho_{\mathrm{w}} - \rho_{\mathrm{g}})}\right\}^2 \text{ for water-gas systems}$$

Note that the water, oil, and gas densities are in g/cc, h is in ft, porosity is fractional, and k is in mD.

14.8 SP and core permeability correlations

A seminal paper published by Archie in 1950 showed core and log data gathered in a sand-shale sequence from an Eocene–Wilcox sandstone in Mercy Texas. Archie noticed that there was a fairly good correlation between the SP deflection in millivolts and the core derived permeability, measured in mD. His data are given in the table and graph below (Fig. 14.10).

SP mV	*k* mD
15	0.001
20	0.001
40	0.2
50	0.2
60	0.5
60	0.1
75	0.5
80	0.5
80	4
100	20
100	20
100	80

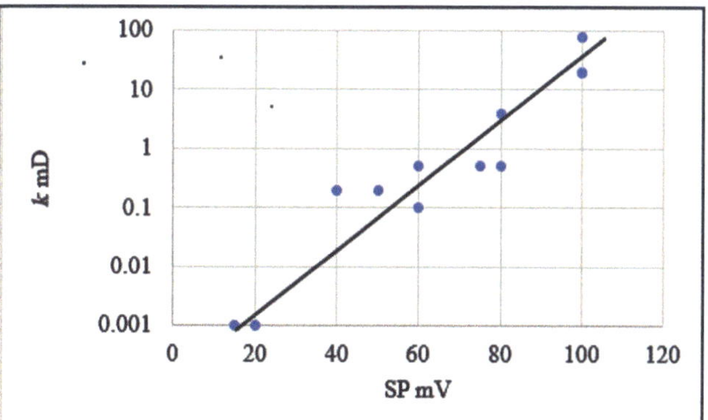

Figure 14.10 SP and permeability correlation.

A "best fit" line of the data points gives an expression for permeability in terms of SP deflection *for this particular dataset* as:

$$k = 10^{\left(\left\{\frac{SP}{18}\right\}-4\right)}$$

While this fit worked well for Archie in Mercy Texas, it may not be applicable elsewhere, and the reader is urged to build custom correlations where core permeabilities and SP logs are available to suite particular plays of interest.

14.9 Summary

There are a number of relatively simple methods for determine permeability from pre-digital well logs. The required inputs can be as simple as a resistivity log or may require additional porosity information coupled with some knowledge of reservoir fluids. The reader is referred to Chapter 3 for help on resistivity logs and to Section III for pointers on porosity estimation from Chapters 4 through 9.

Further reading

Carman, P.C., 1938. The determination of the specific surface of powders. J. Soc. Chem. Ind. Trans. 57, 225.

Carman, P.C., 1956. Flow of Gases Through Porous Media. Butterworths Scientific Publications, London.

Coates, G.R., Dumanoir, J. L., 1973. A new approach to improved log-derived permeability. SPWLA Fourteenth Annual Logging Symposium, 6–9 May.

Kozeny, J., 1927. Ueber kapillare Leitung des Wassers im Boden. Wien, Akad. Wiss. 136 (2a), 271.

Morris, R.L., Biggs, W.P., 1967. Using log-derived values of water saturation and porosity. Transactions of the SPWLA 8th Annual Logging Symposium.

Raymer, L.L., Freeman, P.M., 1984. In-situ determination of capillary pressure, pore throat size and distribution, and permeability from wireline data. SPWLA Twenty-Fifth Annual Logging Symposium, 10–13 June.

Rose, W., Wyllie, M.R.J., 1949. A note on the theoretical description of wetting liquid relative permeability data. Pet. Trans. AIME, 1 (9).

Timur, A., 1968. An investigation of permeability, porosity, & residual water saturation relationships for sandstone reservoirs. Log Analyst IX (4), SPWLA-1968-vIXn4a2.

Wyllie, M.R.J., Gregory, A.R., 1955. Fluid flow through unconsolidated porous aggregates, effects of porosity and particle shape on kozeny-carman constant. Indus, Eng. Chem. 47, 1379–1388.

Wyllie, M.R.J., Rose, W.D., 1950. Application of the Kozeny equation to consolidated porous media. Nature 165, 972.

Wyllie, M.R.J., Splanger, M.B., 1952. Application of electrical resistivity measurements to problem of fluid flow in porous media. Bull. Am. Assoc. Pet. Geol. 36, 359–403.

SECTION VI

Lithology

CHAPTER 15
Lithology indicators

Abstract

Rock and mineral typing today can rely on technologies that did not exist in the pre-digital era. In particular, the spectrographic methods in use today were barely in the experimental stages then and the spectral gamma ray log was only introduced at the end of the pre-digital era. In place of these modern conveniences, which we all now take for granted, the analysts using pre-digital logs will find that basic lithological indicators are just old friends that are still in use today. Here we give them an opportunity to remind us of our ABCs when it comes to figuring out what those F R D Ks[1] are.

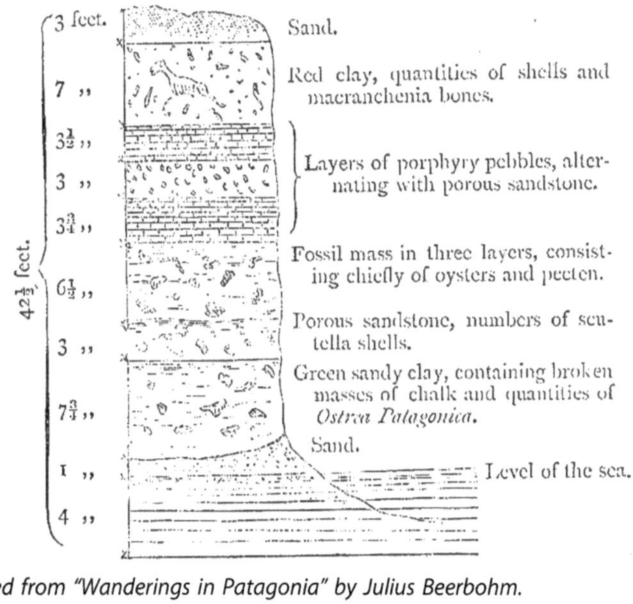

Reproduced from "Wanderings in Patagonia" by Julius Beerbohm.

[1] Funny Rock Don't Know

Formation Evaluation with Pre-Digital Well Logs
DOI: https://doi.org/10.1016/B978-0-12-820232-6.00015-7

Rock typing in the pre-digital era was based on natural gamma ray, neutron, density, and sonic logs. The spectral gamma ray tools were not widely in use until 1969. The Pe logs, based on photoelectric interactions, came later in 1973. The sonic tools capable of discerning both shear and compressional waves did not arrive in the market until 1998. Thus, today's analysts working with pre-digital logs will be obliged to do all sleuthing based on five indicators: spontaneous potential (SP), GR, ϕ_N, ρ_B, and Δt.

15.1 Spontaneous potential

One of the earliest continuously recorded logging parameters was the spontaneous potential, abbreviated SP. Initially it was discovered as an annoying minor offsetting potential that appeared in the borehole and "messed up" the measurement of the normal, inverse and/or lateral electric logs. Its value and importance were soon recognized since the SP potentials clearly indicated where the porous and permeable formations lay when surrounded by impervious formations—such as shales. Early logs even had the SP track labeled as "porosity" (see Chapter 9).

The electrochemical origins of the SP (liquid junction and membrane potentials) are dealt with elsewhere[2] but there is no harm in reminding the reader that the SP can only be recorded in water-based mud systems and the contrast between shales and sands mirrors the contrast between the resistivities of the mud filtrate (R_{mf}) and the connate water (R_w). Fig. 15.1 depicts a standard SP profile in a sand/shale sequence.

The behavior of the SP in low porosity, high resistivity formations is markedly different since the SP currents are restricted by the resistive formations (low porosity limes and dolomites) to flow in the borehole itself and produce "straight line" segments joined by short concave and convex

[2] See Bateman.

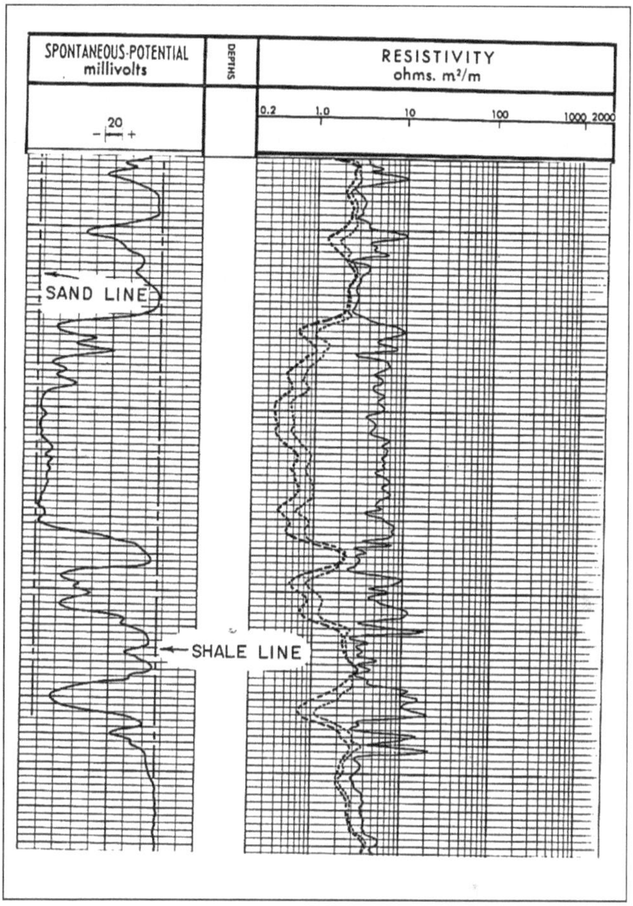

Figure 15.1 SP profile in sand/shale sequence. *Image courtesy Schlumberger.*

segments indicative of sands and shales. Fig. 15.2 gives an example of this behavior.

15.2 Gamma ray log

Radioactive materials have three modes of decay as illustrated in Fig. 15.3. Alpha particles are helium nuclei which can easily be blocked by as little as a sheet of paper. Beta particles are electrons produced when a neutron decays to a positively charged proton. Electrons can be blocked by as little as a sheet of aluminum foil. Gamma rays are less easily blocked and penetrate rocks and steel casing which makes them useful to the log analyst.

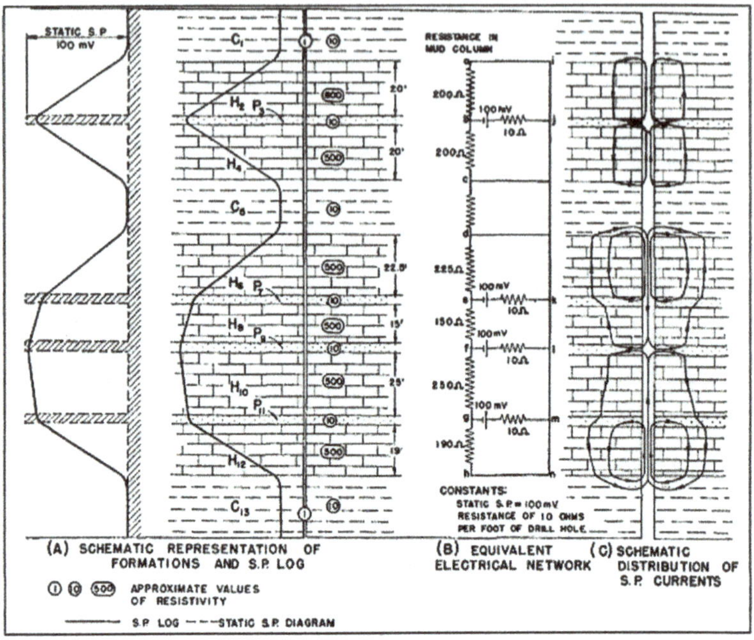

Figure 15.2 SP traces in hard rock—*after H.G. Doll 1948.*

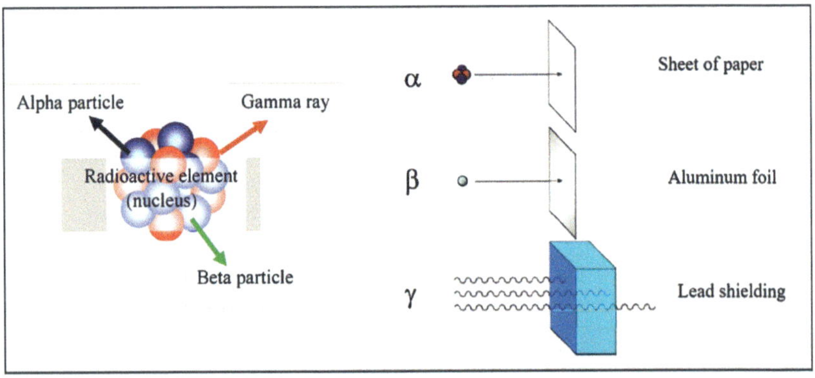

Figure 15.3 Radioactive decay and penetrating power of α and β particles and γ rays.

The early GR logs employed Geiger Mueller-type detectors that employed a cylindrical metal container with a central wire held at a voltage difference with respect to the outer container (see Fig. 15.4). A gamma ray passing through the detector produced ionization and an electrical discharge resulted, which could be counted.

Later in the pre-digital era scintillation counters were introduced which were more sensitive and used sodium iodide crystals. An NaI

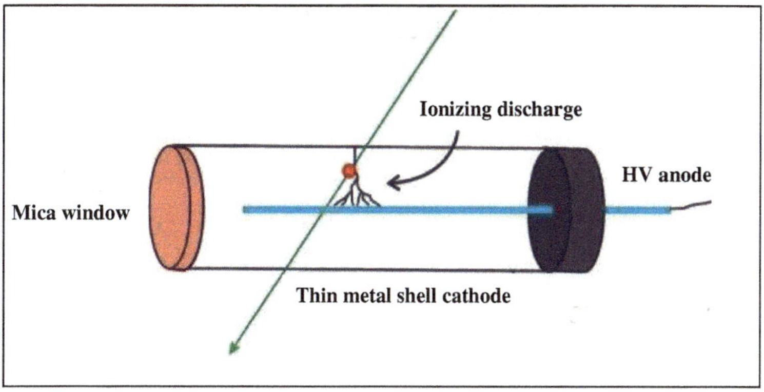

Figure 15.4 Early gamma ray detecting apparatus for wireline logging.

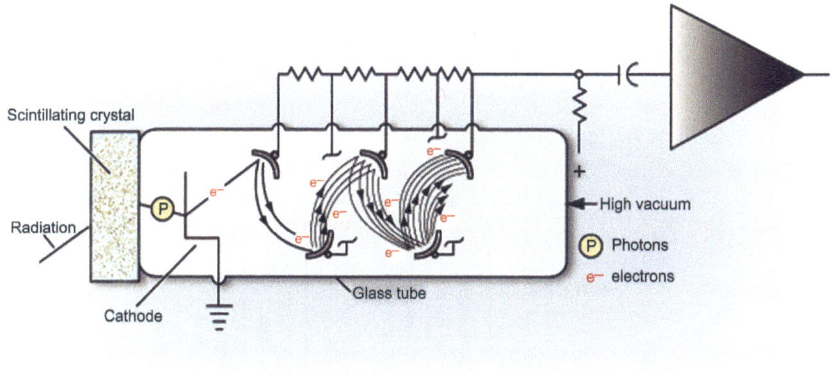

Figure 15.5 A scintillation detector for gamma ray logging.

crystal emits a single photon of light when struck by a gamma ray. This tiny flash of light then strikes a photocathode made of cesium-antimony or silver-magnesium. Each photon, when it hits the photocathode, releases a bundle of electrons. These in turn are accelerated in an electrical field to strike another electrode, producing an even bigger shower of electrons. This process is repeated through several stages until a final electrode conducts a small current through a measurement resistor to give a voltage pulse that signals that a gamma ray has struck the sodium iodide crystal. The system has a very short "deadtime" and can register many counts per second without becoming swamped by too many signals. Fig. 15.5 illustrates a scintillation detector.

Early gamma ray logs were scaled in a variety of units depending on the logging company making the measurement. Fig. 15.6 exhibits a

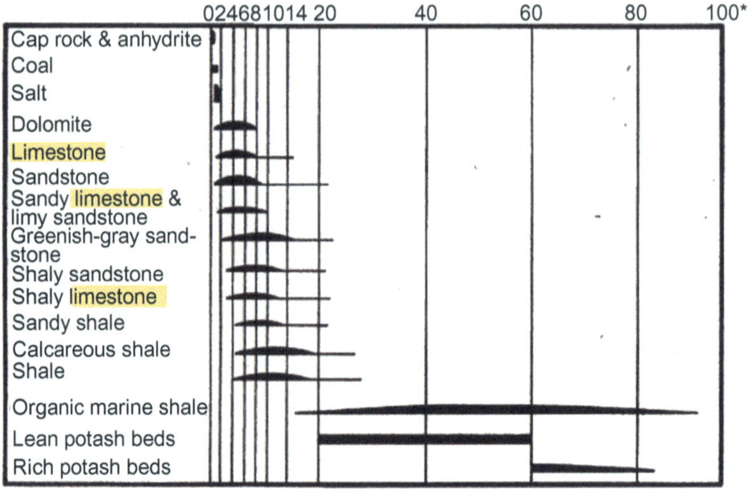

Figure 15.6 Gamma ray response to rocks. – *After Fearon and Mardock AAPG, vol 25. No 9, Sept 1941 P1775.*

Table 15.1 Gamma ray log conversion chart.

TABLE V
GAMMA RAY LOG CONVERSION CHART

Multiply deflection obtained from this→ gamma ray log / To obtain deflection for this log ↓ (By)	Schlumberger Micrograms Ra-eq./ton	McCullough Micro roentgens/hour	Lane-Wells "300" Counts/minute	Lane-Wells "200" Standard units	Halliburton Counts/second	PGA Micro roentgens/hour
Schlumberger Micrograms Ra-eq./ton	1	1.28	0.003	32.5	0.169	1.02
McCullough Micro roentgens/hour	0.783	1	0.0024	25.3	0.133	0.80
Lane-Wells "300" Counts/minute	331	424	1	10,800	55.2	263
Lane-Wells "200" Standard units	0.031	0.405	9.3×10^{-5}	1	0.0048	0.032
Halliburton Counts/second	5.9	7.5	0.018	208	1	6.58
PGA Micro roentgens/hour	0.98	1.25	0.003	31.7	0.152	1

figure from an AAPG paper published in 1941. It shows with remarkable clarity the ranges of gamma radioactivity to be expected for a wide variety of rock types. Of interest too are the units of measurement: "10^{-12} g radium equivalent per gram of rock." Many other scaling units were in use at the time as documented in Table 15.1.

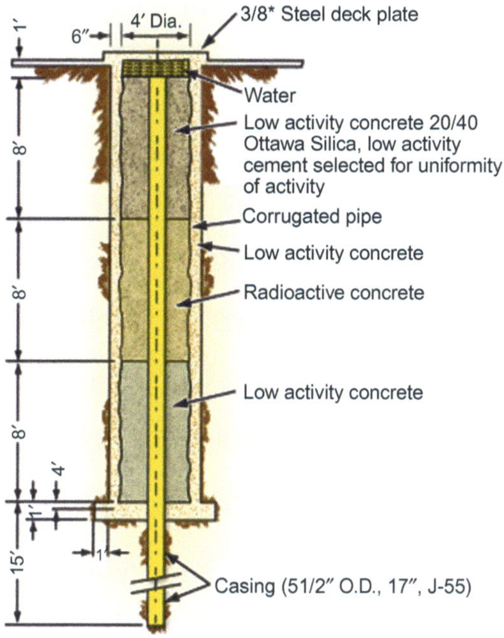

Figure 15.7 API test well for gamma ray calibration.

As can be seen from Table 15. I, the scaling of early GR logs was highly dependent on the logging company used to record the log and intercomparisons of two wells in the same field, logged by different providers, was very difficult. Accordingly, a standard calibration process was agreed upon and, under the aegis of the American Petroleum Institute (API) a test well was constructed on the grounds of the University of Houston which defined the *Gamma Ray API Unit.* Fig. 15.7 details the construction of the well[3] into which any wireline logging company could run their GR logging tool and the difference in the response between the artificial "sand" and the "shale" (radioactive concrete) was, by definition, equal to 200 API units. Thereafter

[3] The original well fell victim to "progress" and was decommissioned in 2012. GR calibration is now taken care of by software running a simulation program.

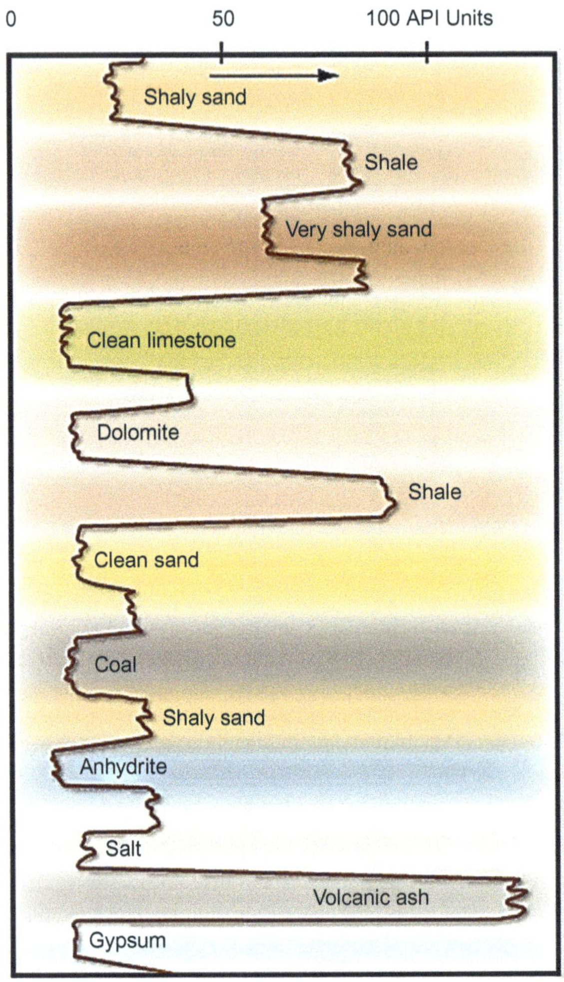

Figure 15.8 Lithology defined by gamma ray API units.

commonly occurring rock types could be categorized by their API units as summarized in Fig. 15.8.

Given an industry-wise standard for ensuring that tools from different logging companies using different sized radiation detectors and counters would now give the same gamma ray API reading in the same formation there still remained the perturbing effects of the hole size, mud weight, and casing size and weight, if present. Today these environmental effects are neatly taken care of "behind the scenes" and the analyst need not be concerned about making any corrections to raw GR measurements. However, in the pre-digital era those corrections had to be made by hand using charts prepared by the logging companies. The appendix to this chapter includes the Schlumberger gamma ray correction charts for both open and cased hole. The reader can get a "feel" for the magnitude of the corrections by considering that, other things being equal:

- changing hole size from 8½″ to 9 5/8″ produces an observed GR reduction of ≈ 20%
- changing mud weight from 8 to 12 lb/gal produces a reduction of ≈ 40%
- changing from open hole to cased hole with 7″ casing produces a reduction of ≈ 60%

Although substantial, these environmental effects need not overly affect log analysis if only relative GR deflections are considered. Calculations of V-shale are unaffected by the absolute GR API units involved once "clean" and "shale" readings are established.

Worthy of note is the advent of the spectral gamma ray logging tools that came on the scene in the late 1960s. These required some considerable computational horsepower in the surface units of the logging company offering the service and so strictly speaking cannot be included in discussions of pre-digital matters.

15.3 Combinations by pairs of sensors

To determine porosity from neutron, density or sonic logs the nature of the rock matrix must be known. Without the help from modern sensors, such as Pe, the analyst of yesteryear had to resort to using *pairs* of sensors to eliminate the unknown quantity. We will investigate these plots and list their modes of use along with their good and bad points.

The first neutron-density plots were based on the old "count rate" neutron logs (see Chapter 5) and their output of a "limestone porosity index" was matched up with a formation density, in gm/cc, derived from a pioneering gamma–gamma (uncompensated) density tool (see Chapter 6). Fig. 15.9 gives an example of such a plot from the mid-1960s. In many aspects it will look familiar to today's analyst in as much as the three main lithologies (sandstone, limestone, and dolomite) appear as lines marked off with porosity ticks. The curved nature of the sand and

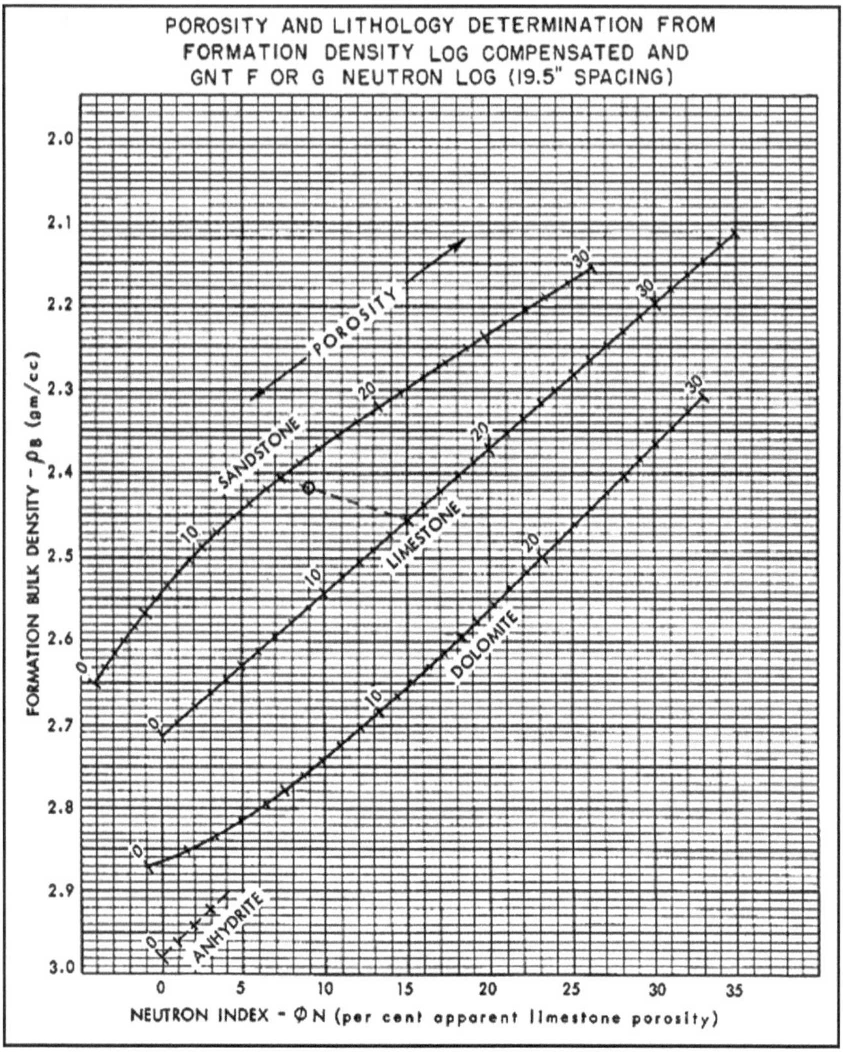

Figure 15.9 Early neutron-density crossplot. *Image Courtesy Schlumberger.*

dolomite lines reflects the underlying physics and chemistry of neutron logging, common to all the later tools as well.

A second combination was the neutron-sonic plot, as shown in Fig. 15.10. In cases of mixed lithology plotted points might fall between lines for sandstone, limestone or dolomite but usually a fair estimate of porosity could be made by interpolation.

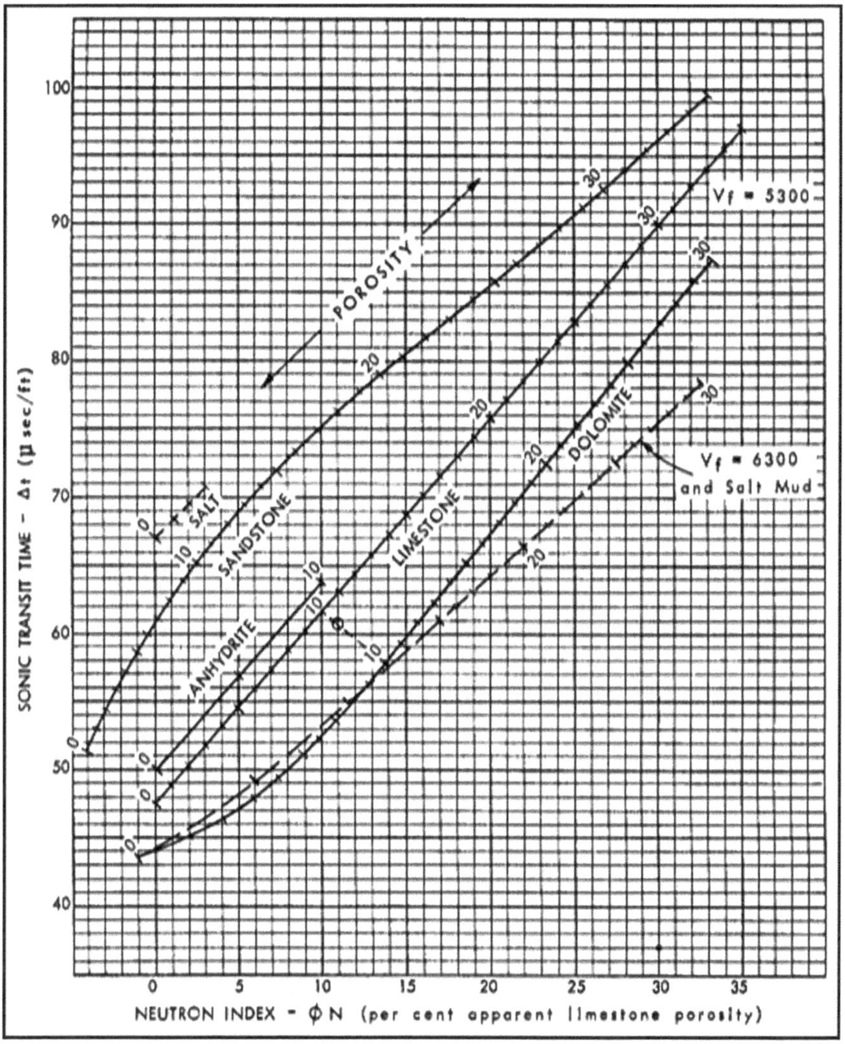

Figure 15.10 Early neutron-sonic crossplot. *Image Courtesy Schlumberger.*

The third plot in this cycle is the density-sonic plot a shown in Fig. 15.11. It is noteworthy for the positioning of the zero porosity minerals that appear as points only.

Apart from the plots involving *pairs* of porosity log readings another set of plots attempted to simultaneously include the information from *all three logging* devices on a single plot that could uniquely define the matrix of the rocks logged through. The "M and N" plot essentially took a

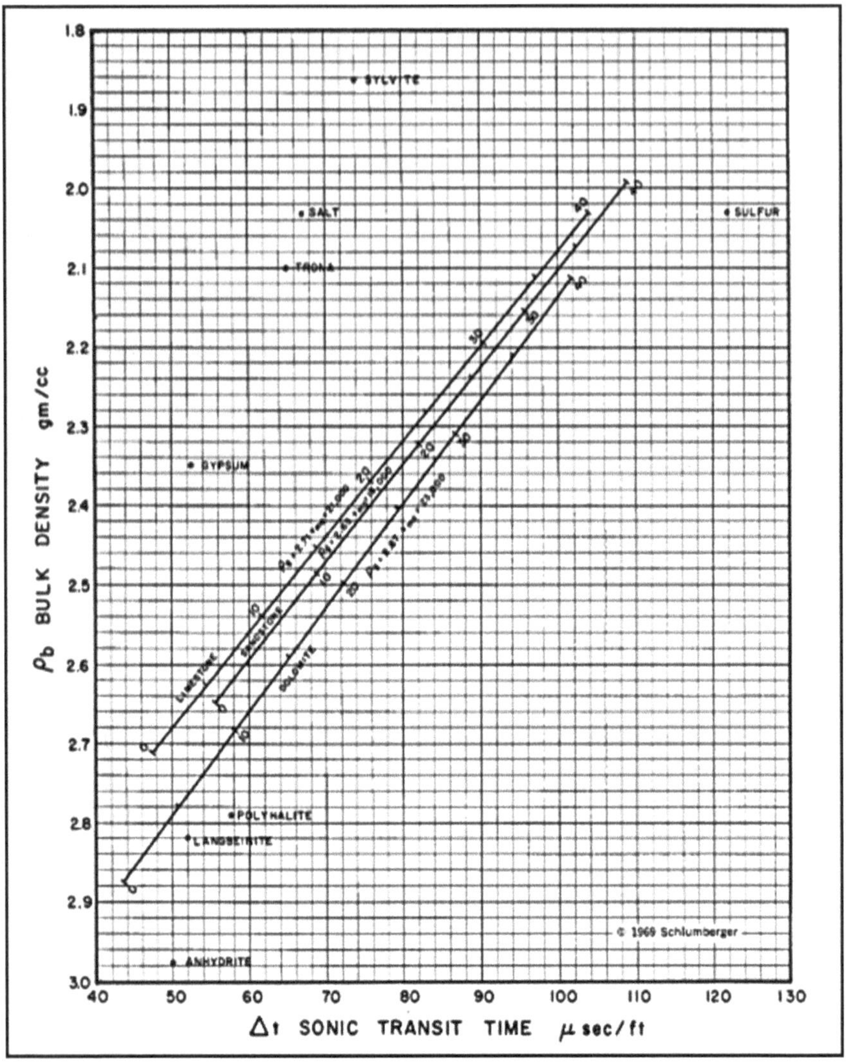

Figure 15.11 Early sonic-density plot. *Image Courtesy Schlumberger.*

density-sonic crossplot (Fig. 15.11) and calculated the slope of a line that joined the 100% water-bearing point to the plotted point of Δ_t and ρ_B. This was given the symbol "*M*."

$$M = \frac{\Delta t_{fl} - \Delta_t}{\rho_B - \rho_{fl}} x \frac{1}{100}$$

Then, on a density-neutron plot (Fig. 15.9) one could calculate the slope of a line connecting the 100% water-bearing point to the logged point. This was given the symbol "*N*."

$$N = \frac{\phi_{\mathrm{Nfl}} - \phi_{\mathrm{N}}}{\rho_{\mathrm{B}} - \rho_{\mathrm{fl}}}$$

The values of *M* and *N* were then plotted on the *M* versus *N* chart of Fig. 15.12. Alternatively, the table, shown in Table 15.2, could be consulted:

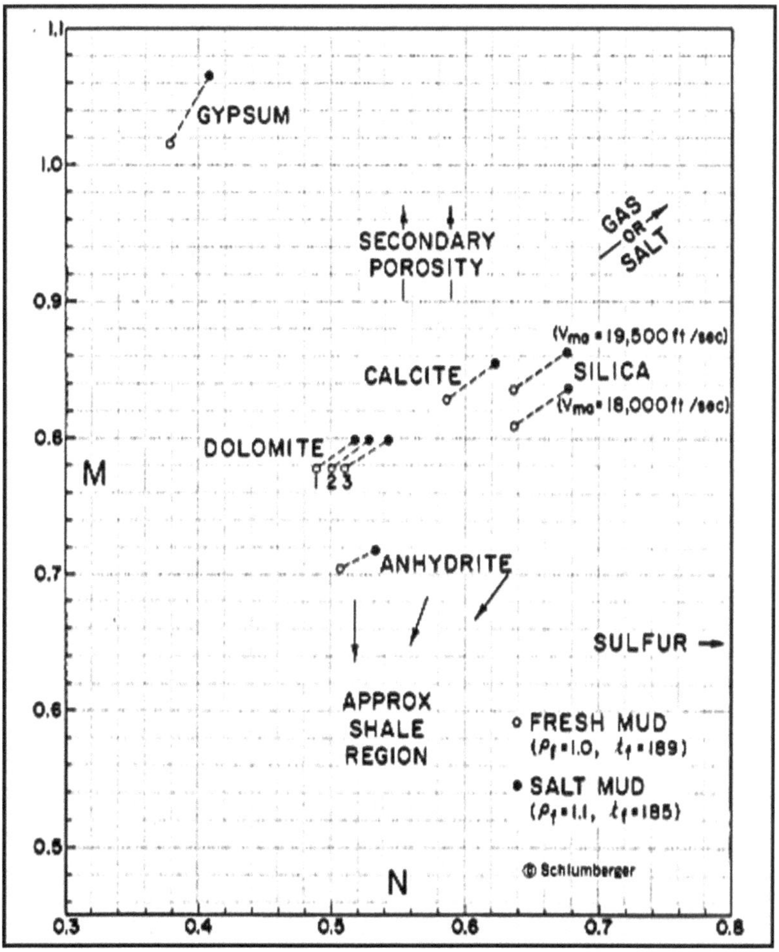

Figure 15.12 *"M and N"* plots for mineral identification. *Image Courtesy Schlumberger.*

Table 15.2 "*M*" and "*N*" values for reservoir rocks and minerals.

Mineral	Fresh Mud (ρ_f=1)		Salt Mud (ρ_f=1.1)	
	M	N	M	N
Sandstone	0.81 to 0.835	0.628	0.835to 0.862	0.669
Limestone	0.827	0.585	0.854	0.621
Dolomite	0.778	0.516 to 0.532	0.8	0.544 to 0.561
Anhydrite	0.702	0.505	0.718	0.532
Gysum	1.015	0.378	1.064	0.408

Table 15.3 Rho matrix and delta-T matrix for reservoir rocks and minerals.

Mineral	Δt_{ma}	ρ_{ma}
Sandstone	51.2 to 55.5	2.65
Limestone	47.5	2.71
Dolomite	43.5	2.87
Anhydrite	50	2.98
Gysum	52	2.39
Salt	67	2.03

The shortcomings of the plot were that, depending on the salinity of the mud system, the position of the 100% water-bearing point changed and so did the values of *M* and *N*, leaving the plot with uncertainties about what it was actually telling the analysts about the rock matrix.

A similar plot that followed later was based on a pair of plots that ignored porosity and indicated two parameters that were direct indicators of the rock type—namely the sonic travel time in the matrix (Δt_{ma}) and the matrix density (ρ_{ma}). These were then combined in a final plot called the matrix identification plot or MID plot. The values for these two parameters are listed below in Table 15.3.

Fig. 15.13 gives an example of the neutron-sonic plot that was gridded with only lines of constant matrix travel time, or delta-T matrix, given

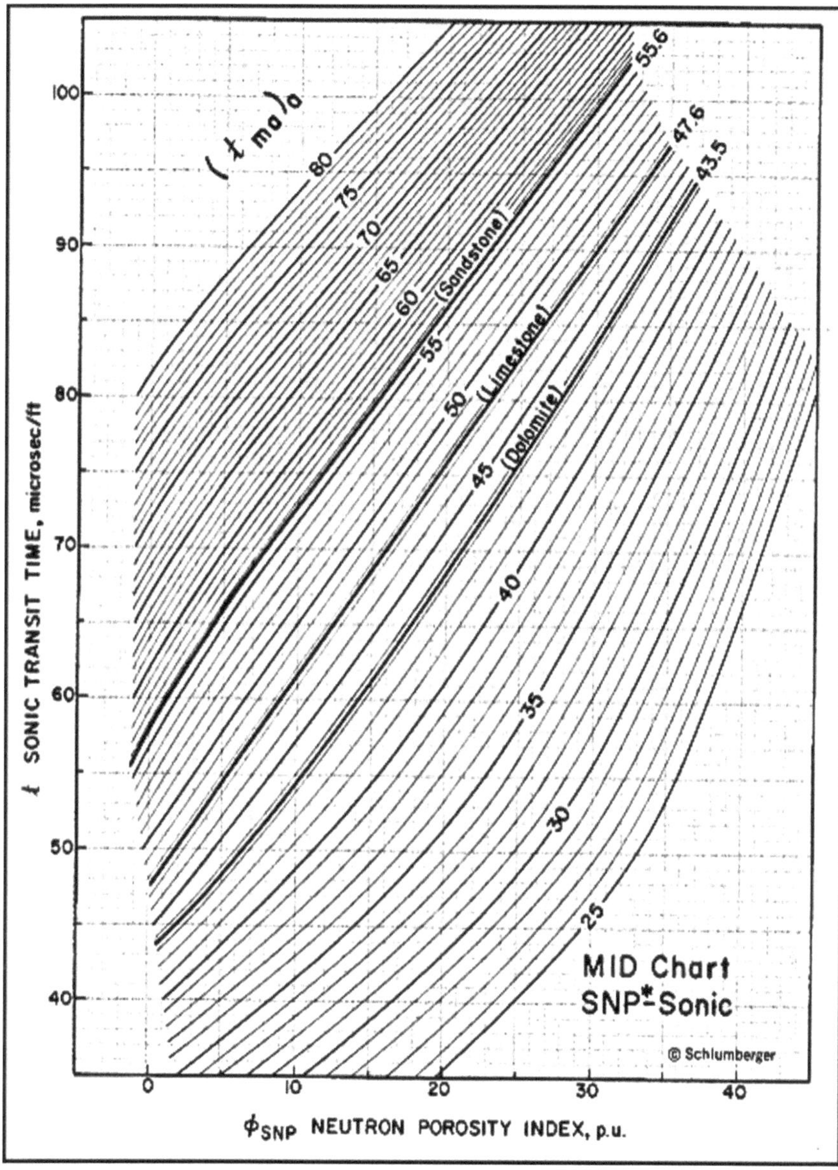

Figure 15.13 Neutron-sonic plot for delta-T matrix. *Image Courtesy Schlumberger.*

the symbol Δt_{ma}. Note that this version of the plot used the sidewall neutron porosity (SNP) readings[4]. Fig. 15.14 shows the neutron-density part of the MID plot.

[4] More familiar to today's readers is the version of this plot with the source of the neutron information as a compensated neutron log (CNL) which is included as Figure 15.A.3 in the Appendix to this chapter.

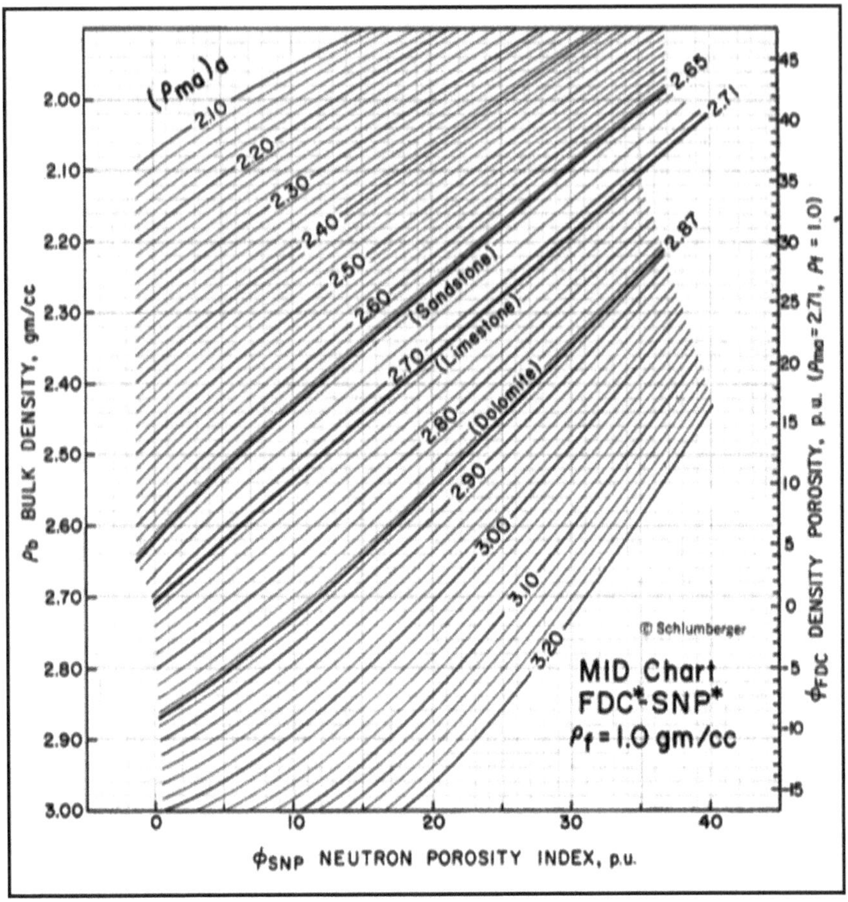

Figure 15.14 Neutron-density plot for rho matrix. *Image Courtesy Schlumberger.*

The product of the neutron-density and neutron-sonic plots are then entered on the final plot to determine the matrix material logged. Fig. 15.15 shows this final "answer" plot.

15.4 Combinations of porosity and resistivity

The combination of resistivity and porosity on the Hingle plot has been covered in Chapter 11. In that context it was used as a means of determining the value of aR_w. However, it can equally be used to determine the matrix point on any porosity measuring device.

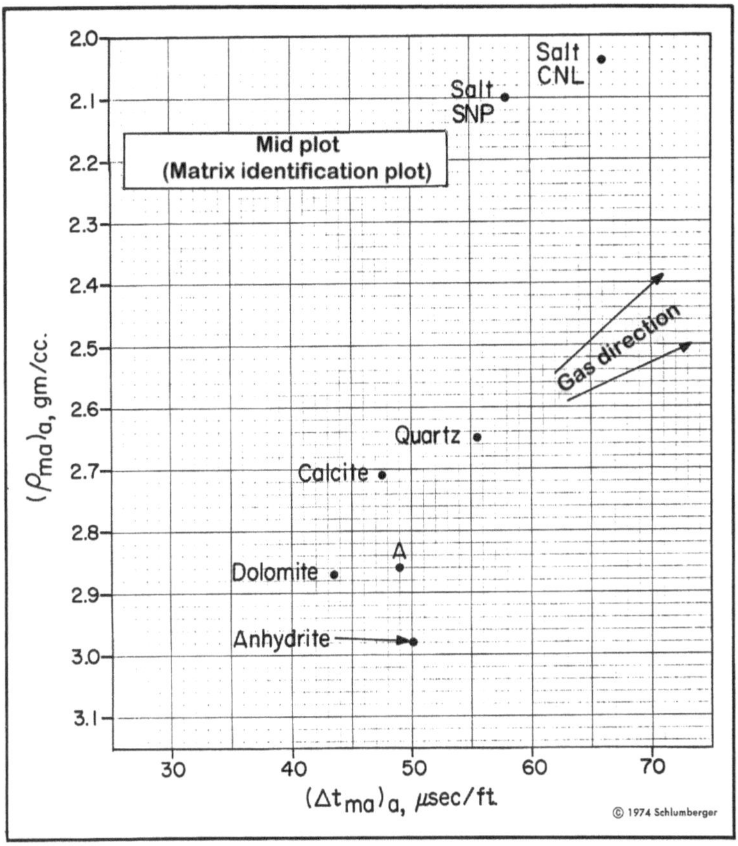

Figure 15.15 The matrix identification plot (ρ_{ma} vs Δt_{ma}). *Image Courtesy Schlumberger.*

As before, the function of resistivity used is the reciprocal root of R_t (RRRT). Various versions of the resistivity grid scale are available in chart books. The plot of Fig. 15.16 was constructed to accommodate the formation factor to porosity relation of

$$F = \frac{0.62}{\phi^{2.15}}$$

Against this function of formation resistivity may be plotted the one hundred percent water-bearing readings from either a sonic or density or neutron log on a linear scale. The resulting plot will reveal a trend from the top right to the bottom left of the chart area which will intersect the infinite resistivity

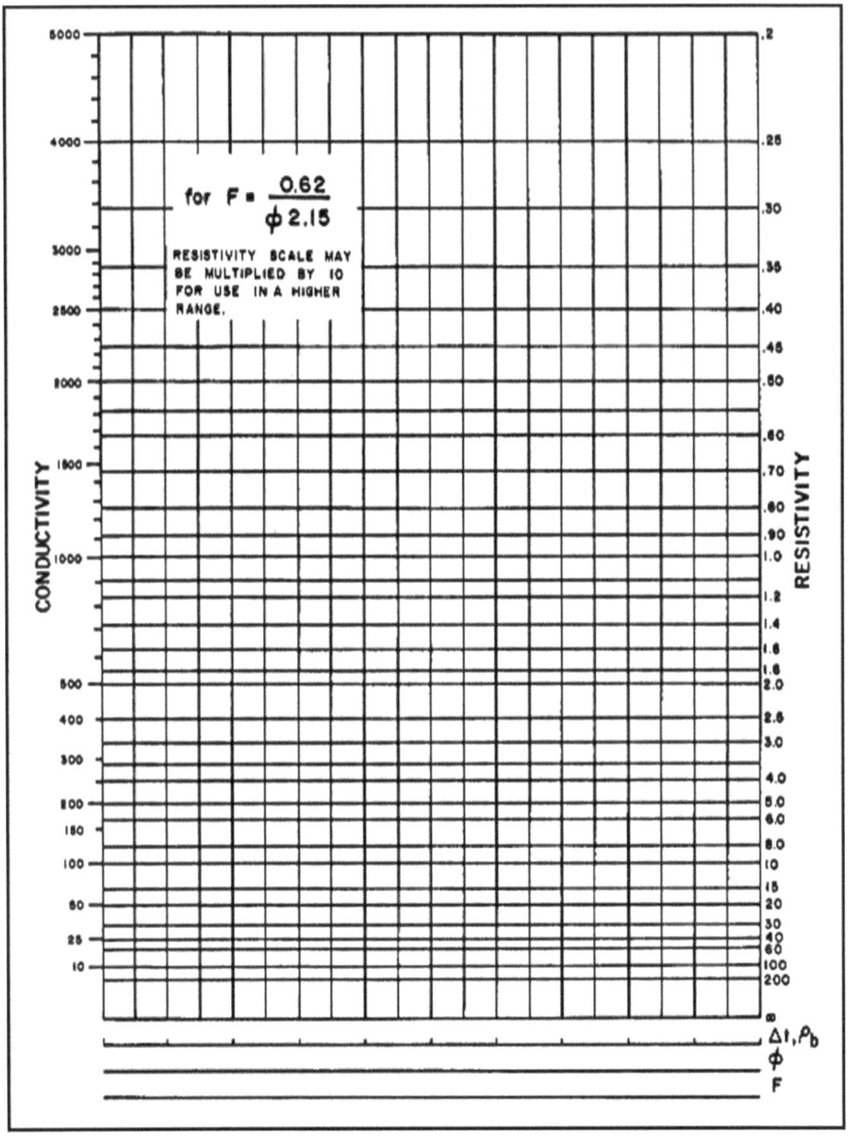

Figure 15.16 Hingle plot for determining matrix point. *Image Courtesy Schlumberger.*

line at precisely the zero porosity matrix point either in neutron porosity units, matrix density in gm/cc or sonic travel time in μseconds/ft.

Today's log interpretation software packages prepare continuous curve plots of these lithological indicators which allow the analysts to rapidly identify formation tops and mineralogy. The analyst working with pre-digital tools was obliged to manually read points off the available logs, plot them by hand on the appropriate charts, read values of apparent matrix density and/or travel time and then replot on a second chart to get the required information on rock type. Working with old logs today still offers the same challenge. It is, of course, possible to digitize old logs from paper prints and then proceed with a computerized interpretation routine. Analysis of the time and effort involved suggests that, for most of the cases, hand picking and plotting of carefully selected points, chosen with the benefit of a little hindsight, can produce effective results of comparable utility.

By way of summary, most pre-digital log suites will have at least an SP and a gamma ray to work with on formation rock typing. If, in addition, there are at least two of the three "porosity" logs (sonic, neutron, and density) then the old log analysts (who could be young!) should be able to tie down lithology with confidence without the modern benefits of hi-tech post-analog logging sensors.

Further reading

Burke, J.A., Campbell Jr., R.L., Schmidt, A.W., 1969. The litho-porosity crossplot. The Log Analyst .

Edmundson, H., Raymer, L.L., 1979. Radioactive logging parameters for common minerals. Log Analyst .

Howell, L.G., Frosch, Alex, 1939. Gamma ray well logging. Geophysics IV (2), 106–114 (1939).

Hughes, J.D. "Radioactivity logging provides valuable data for gulf coast operations"; Tomorrow's Tools — Today, 9, 4, p. 4.

Poupon, A., Hoyle, W.R., Schmidt, A.W., 1971. Log analysis in formations with complex lithologies. J. Pet. Tech 23 (8).

Raymer, L.L. and Biggs, W.P., 1963. Matrix characteristics defined by porosity computations. SPWLA Symposium.

Appendix to Chapter 15

Figs. 15.A.1–15.A.3

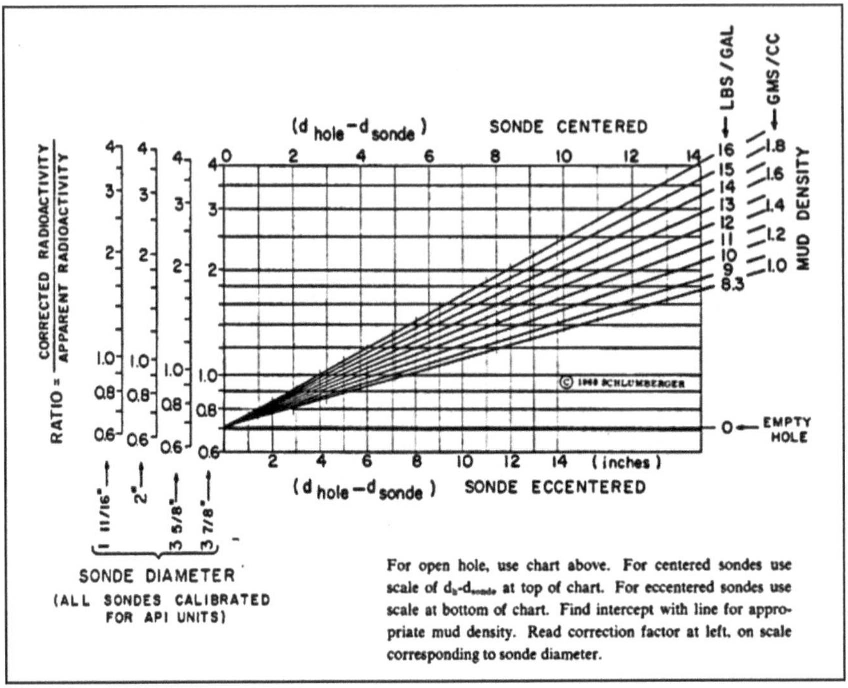

Figure 15.A.1 Gamma ray correction chart for open hole. *Image Courtesy Schlumberger.*

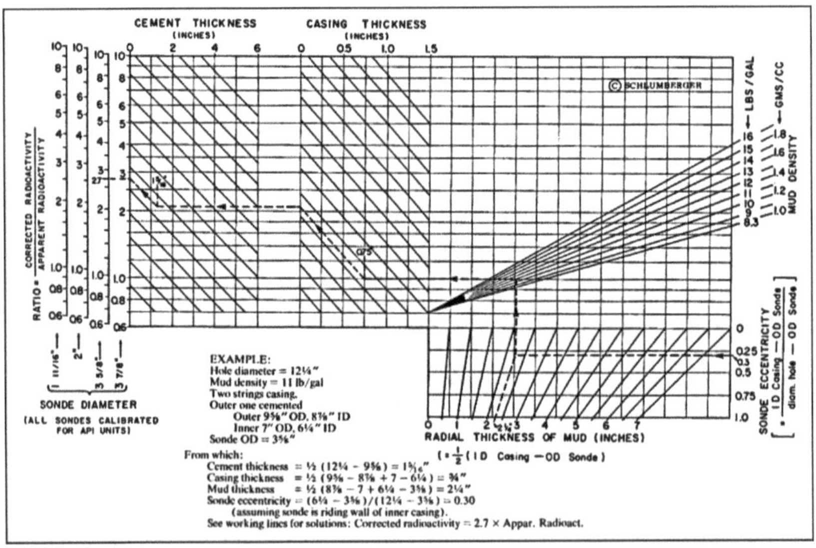

Figure 15.A.2 Gamma ray correction chart for cased hole. *Image Courtesy Schlumberger.*

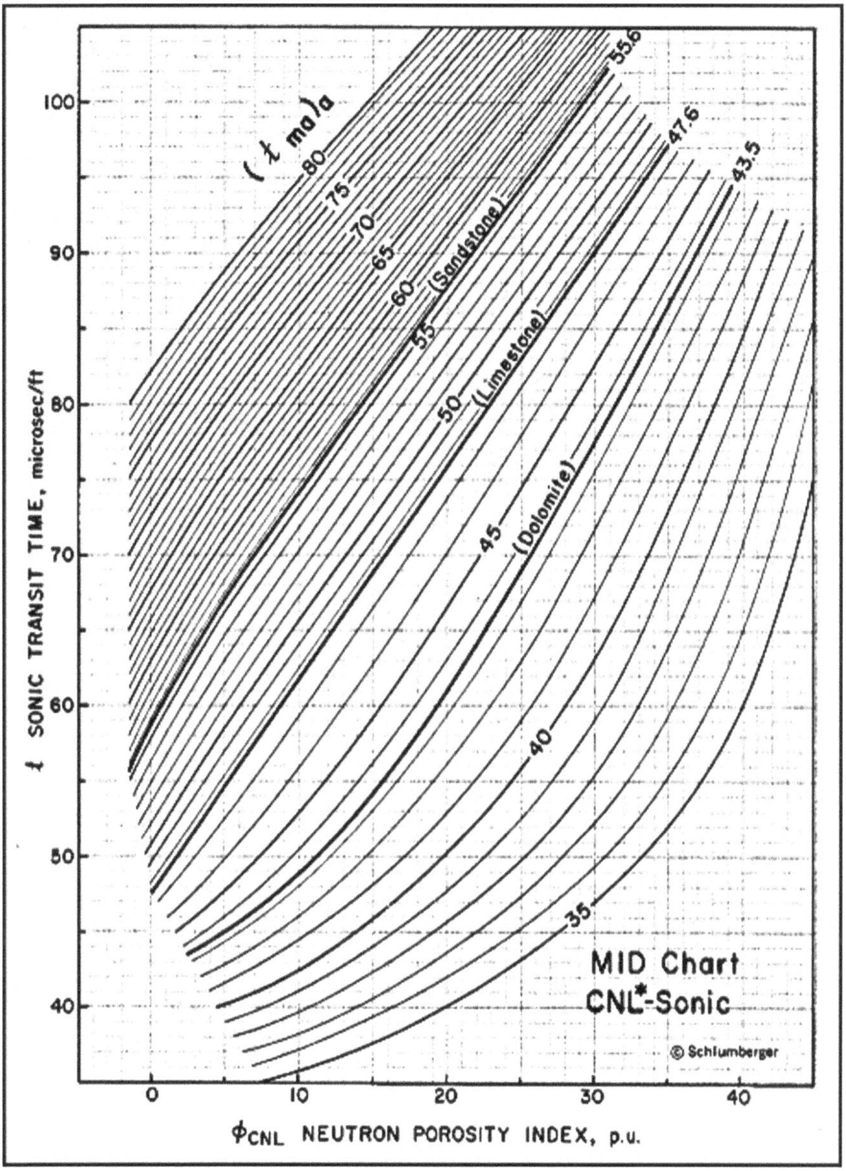

Figure 15.A.3 Neutron-sonic plot for Δt_{ma} using CNL neutron. *Image Courtesy Schlumberger.*

SECTION VII

Miscellaneous

CHAPTER 16

Dipmeter

Abstract

Dipmeter logging has a long and eventful history and has witnessed many of the most ingenious advances to the science and art of determining the dip and azimuth of subsurface formations. Along the way, the pursuit of structural and sedimentary detail leads to the modern technology of formation imaging, initially by electrical means, as an outgrowth of the ever more sophisticated multiarm, multielectrode dipmeter tools, and then, later, by the use of other means including ultrasonic sensors and the use of radioactivity—both natural and induced.

Henri Doll testing the first Teleclinometer in Baku in 1932. *Image courtesy Schlumberger.*

DOI: https://doi.org/10.1016/B978-0-12-820232-6.00016-9

16.1 The need for dipmeters

Today dipmeter surveys have a variety of applications. At the lowest level, the raw data may be used:

- To compute a deviation survey and true vertical depth
- To compute the integrated hole volume
- As an aide to fracture detection

At a higher level, computed dipmeter results may be used to determine:

- gross geologic structural features crossed by the wellbore,
- sedimentary details within the reservoir (grading, bioturbation, etc.),
- thin bed analysis,
- the depositional environment (paleocurrents, etc.), and
- true stratigraphic thickness and true vertical thickness.

At the highest level, computed dipmeter results from many wells may be combined to produce:

- structural cross-sections,
- trend surface maps, and
- detailed views of rock texture.

Thus, the most important applications of the dipmeter survey are in exploration drilling to help identify local structure and stratigraphy and in development drilling to help map the productive horizons and indicate directions to follow for further field development.

The first attempts relied on the cutting and retrieval of oriented cores. However, the method proved overly complex and cumbersome and did not lend itself to wide coverage in any one well or, indeed, fieldwide. As the industry advanced through the pre-digital age, a number of dipmeter-type devices evolved and are documented here—more out of historical interest than as a guide to the use of what precious few examples of the early logs of this type still exist.

16.2 Anisotropy dipmeter

Where bedding planes are anisotropic, the formation resistivity parallel to the bedding planes is measurably different from the resistivity normal to them. This salient fact is the basis for the modern "3D" induction tools that measure both horizontal and vertical formation resistivities and have

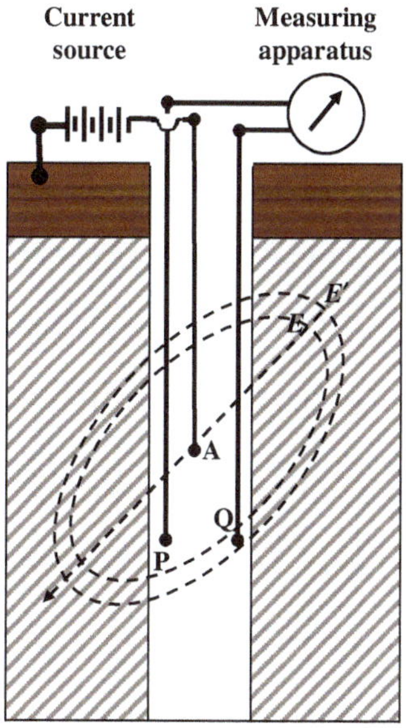

Figure 16.1 The principle of the anisotropy dipmeter.

been so successful in improving log interpretation in laminated sand-shale sequences. It should be no surprise that the earliest dip measuring apparatus relied on precisely the same principle. Fig. 16.1 illustrates that the equipotential surfaces, E and E', due to current flowing from electrode A in the hole, would thus be ellipsoids with their long axis parallel to the bedding plane. Consequently, potential measuring electrodes P and Q on opposite sides of the borehole detect slightly different potentials and, if both are rotated about the vertical axis of the borehole, will reach a maximum potential difference when aligned with the direction of the dipping strata.

Physical rotation of the potential measuring electrodes was avoided in practice using four measure electrodes arranged at 90 degrees to each other. The orientation of these four electrodes was monitored by a device known as a "teleclinometer." These devices had limited applications since they worked only where there were thick, conductive, homogeneous, and anisotropic bedding planes, such as shales. Moreover, while they indicated the direction of a dipping bed, they did not give any reliable quantitative measure of the bed dip angle per se.

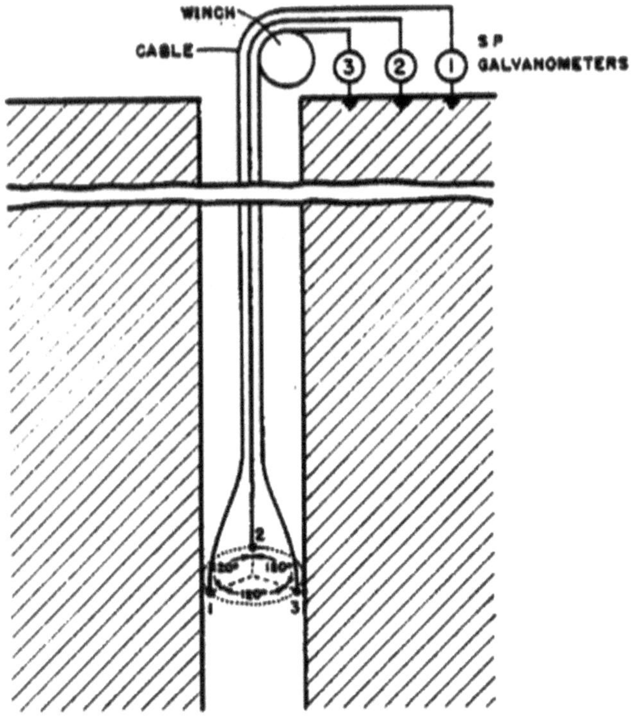

Figure 16.2 SP dipmeter apparatus.

16.3 SP dipmeter

In 1942 the SP dipmeter was introduced. A simple arrangement of three SP-measuring electrodes spaced 120 degrees apart and positioned close to the wall of the borehole produced three SP traces displaced in depth anywhere that the bedding plane (drilled though and logged) was not horizontal (assuming a vertical well bore). Fig. 16.2 illustrates the circuit arrangement for the three electrodes.

The SP dipmeter could provide a continuous log as shown in Fig. 16.3, which illustrates a recording over some 15 ft of the wellbore. Slight displacements between the "peaks and valleys" of the three SP curves reflect the bed dips. The orientation of the dipping planes was deduced by stationary measurement by a "photoclinometer" that

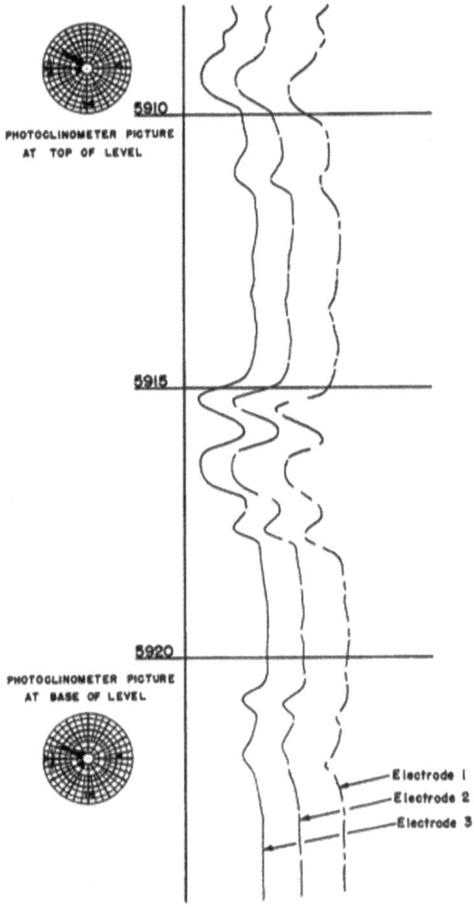

Figure 16.3 SP dipmeter log recording.

incorporated a magnetic compass as well as a small ball that was free to roll in a hemispherical "cup" that was ruled with both circles to delineate hole deviation from vertical and azimuth lines to show the hole direction.

Fig. 16.4 shows the photoclinometer. At each "station," a photo was taken to document the tool's electrode orientation and the hole angle and drift. From these, the true formation dip and azimuth were calculated. A worksheet for computing dips from the displacements of the three SP curves is shown in Fig. 16.5.

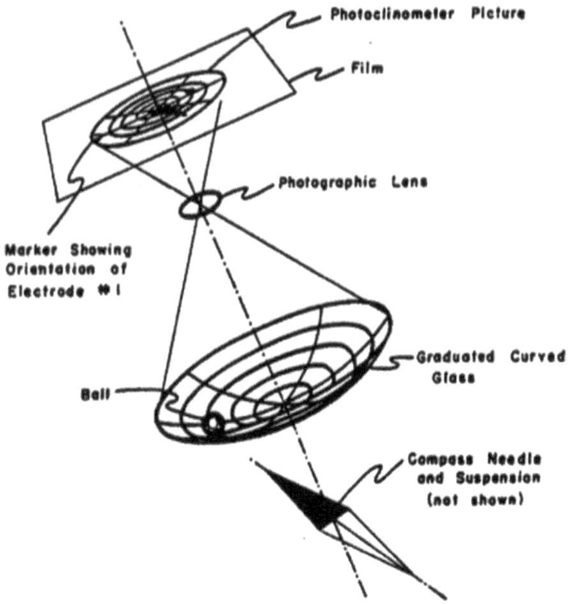

Figure 16.4 SP dipmeter photoclinometer.

STATION	Depth Interval	From Magnetic North			Displacement of Curves in reference to I.		DIP		
		Drift Azimuth	Drift Angle	Orient. No. I	II	III	Dip Angle	Direction from Mag. N.	Direction from True North
A1	2000 to 2020	342	0 45	77	U 0.4	U 1.2	9	120	S 50 E
A2	2004 2024	342	0 45	77	U 0.4	U 1.2	9	120	S 50 E
B1	2856 to 2875	234	2 00	229	D 0.6	D 2.2	17	91	S 79 E
B2		234	2 00	16	D 0.8	U 1.2	15	93	S 77 E
C1	3135 to 3158	241	1 00	119	U 1.8	U 0.6	13	75	N 85 E
C2		241	1 00	299	D 1.9	D 0.9	13	81	S 89 E
D1	3520 to 3535	237	2 00	214	-0-	D 1.8	15	88	S 82 E
D2		237	2 00	75	U 1.2	U 2.0	15	94	S 76 E

Figure 16.5 Worksheet for dip calculation from sp dipmeter log.

16.4 Resistivity dipmeter (three arms)

In 1945 the SP dipmeter was superseded by the three-arm resistivity dipmeter[1]. This employed three articulated arms equipped with "micro" resistivity electrodes that rode along the borehole wall recording minor resistivity variations with depth in accordance with the bedding planes encountered. Fig. 16.6 shows the tool that initially did not have powered

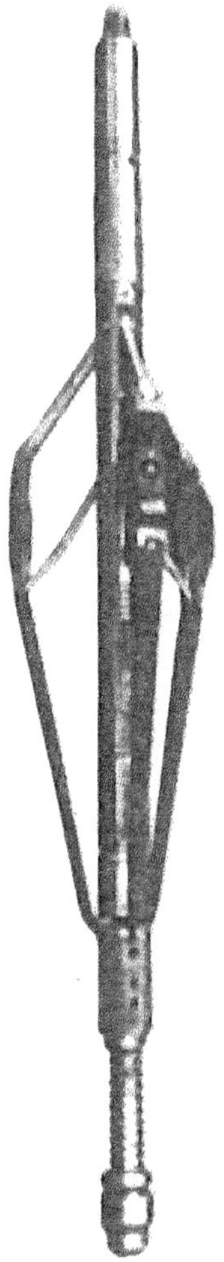

Figure 16.6 Three-arm continuous dipmeter tool. *Image courtesy Schlumberger.*

[1] Initially the tool orientation data were recorded at stations only, and it was not until 1952 that a dipmeter tool with continuous orientation data was available, which eliminated the need to stop and take station measurements.

caliper arms. The three spring-loaded arms were wired in the closed position in order to get the tool to the bottom of the hole to be logged. At that point, the operator had to remotely fire a blasting cap that would hopefully cut the restraining wire that held the arms shut, thus allowing the resistivity electrodes to make contact with the formation.

A typical field log with the three dip curves in tracks 2 and 3 along with a caliper (hole size) recording is shown in Fig. 16.7. Track 1 shows the orientation data with recordings of the azimuth of electrode 1 and the hole deviation from vertical (drift). As can be seen, the hole size is close to 8½″, the dip curves exhibit only small depth displacements between themselves (indicating quasi horizontal bedding), the electrode No. 1 is pointing a little East of North and the hole is deviated 2 degrees to the North-East.

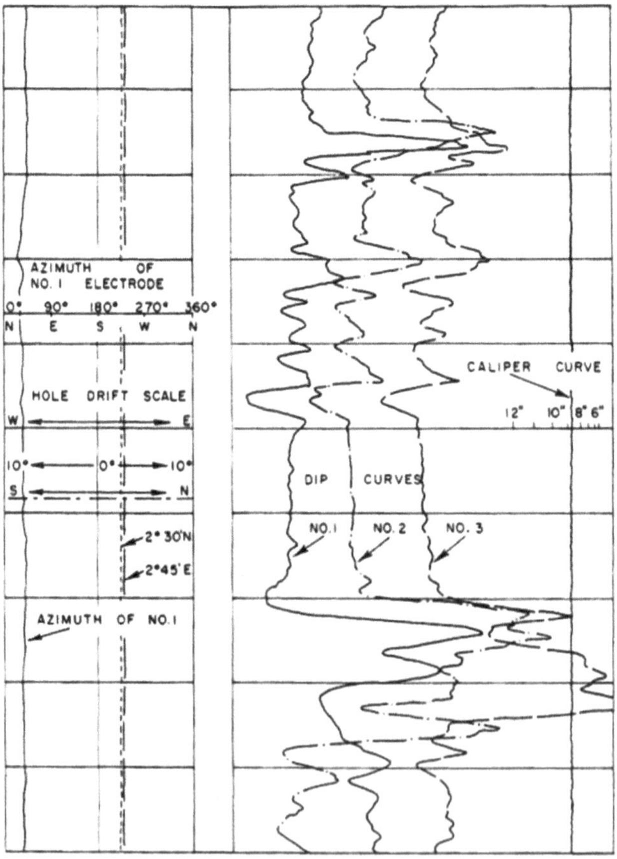

Figure 16.7 Resistivity dipmeter log—three arms.

Figure 16.8 Mechanical model for dipmeter computations. *Image courtesy Schlumberger.*

Dip computations from the three-arm dipmeter were handled by optically correlating the resistivity dip traces to align anomalies on one trace with a similar one on an adjacent trace. A paper copy of the field film record was placed under the film so that any two traces could easily be moved up-hole or down-hole to get as exact a fit to an adjacent dip curve as possible. The displacements in inches (and fractions of an inch) were then coupled with the orientation readings to calculate the dip angle and the direction of downdip (the azimuth). A mechanical model could also be used, as illustrated in Fig. 16.8. The model allowed the analyst to adjust the displacements between the dip traces, the hole size, as read from the caliper, and the compass and hole deviation readings resulting on the production of a tilted plane that gave both the dip angle and the dip azimuth. Of course, this exercise was extremely labor intensive and was soon surpassed using an optical comparator.

The optical comparator offered a more sophisticated method to determine the displacement between adjacent dip traces using half silvered mirrors. The image of a short section of one trace could be optically laid over the actual trace of its adjacent trace and then slid back and forth to determine a best fit and the corresponding displacements. Once so determined, the data were punched onto 80-column IBM cards and fed into a computer where the required trigonometry was applied to spit out the answers, which could then be plotted on an appropriate device. The operator needed to be a skilled dipmeter analysts, and these experts were few and far between. To the knowledge of the present author, the last known optical comparator was for many years owned and operated by Betty Van–Norman on the Louisiana Gulf Coast, but her device fell victim to hurricane Katrina in 2005.

16.5 High-resolution dipmeter (four and six arms)

The four and six-arm dipmeters were entirely digital and recorded all resistivity and orientation data in digital format directly on magnetic computer tape. Improvements on those designs are still in use today, and further discussion here is not relevant. It is suffice to say that the improvements in dipmeter logging, recording, and computation have come from a combination of higher depth sampling rates for the resistivity traces and improved measurements of the tool's spatial geometry.

16.6 Fracture finding

Another application of the dipmeter survey is in the detection of fractures. There are many methods available for fracture detection, and none, on its own, is a completely reliable diagnostic. The dipmeter should be viewed as just one of many methods for fracture finding and should be used to complement other methods. The theory is very simple. An open fracture will be invaded with mud filtrate and will, therefore, offer a less resistive path to electric current. If one of the dipmeter pads happens to lie in front of a fracture, it will record a low resistivity value. Another pad at the same depth may not be in front of a fracture and will record a higher resistivity. Comparison of adjacent pad traces should reveal the presence of a fracture when the two resistivity values are different. Various ways are available to display the curves in order to highlight such differences. Figure 13.15 (see Chapter 13—quick-look methods) shows one such presentation. Note that since the orientation of the dipmeter tool is known, the orientation of the fracture can also be deduced.

Further reading

Boucher, F.G., Hildebrandt, A.B., Hagen, H.B., 1951. New dip-logging method; Second Symposium on Subsurface Geological Techniques. School of Geology, University of Oklahoma, pp. 101–110.

de Chambrier, P., 1953. The microlog continuous dipmeter, Paper Presented Before the Houston Meeting. 23–26 March. Society of Exploration Geophysicists.

Doll, H.G., 1942. The S.P. Dipmeter, A.I.M.E. Technical Publication No. 1547, Austin Meeting.

Gilreath, J.A., 1960. Interpretation of dipmeter surveys in Mississippi. First Annual Meeting of the Society of Professional Well Log Analysts. Tulsa, OK.

Rowland, D.A., 1957. An Analytical Method for Solving Dipmeter Data. Society of Petroleum Engineers.

Schlumberger, C., Schlumberger, M., Doll, H.G., 1933. The Electromagnetic Teleclinometer and Dipmeter. World Petroleum Congress, London.

Schlumberger, C., Schlumberger, M., Doll, H.G., 1935. The Electromagnetic Dipmeter Determination of the Direction of Dip of the Sedimentary Strata Crosscut by Drillings. Int. Congress Mines Metallurgy and Geological Technology, Paris.

Schlumberger, C.M., Leonardon, E., 1934. Some observations concerning electrical measurements in anisotropic media, and their interpretation, p. 159. Electrical coring: a method of determining bottom-hole data by electrical measurements, p. 237. A new contribution to subsurface studies by means of electrical measurements in drill holes, p. 273. Petrol. Trans. AIME 110.

Schlumberger, U.S. Patent 2,176,169 on SP Dipmeter.

CHAPTER 17

Formation tester

Abstract

Wireline formation testing provided a window on the productivity of a potential "pay zone"—identified by fairly basic logging tools. In many cases a "go/no go" decision on casing and cementing a new well was based on what a one-shot wireline formation tester could tell the operator. The industry produced some subtle methods for predicting hydrocarbon production type and water cut from limited recoveries made by these simple (by today's standards) tools. Although it is doubtful that many of these records will show up in the modern analyst's expedition into old filing cabinets it is worthwhile to become familiar with the methods used in pioneering times.

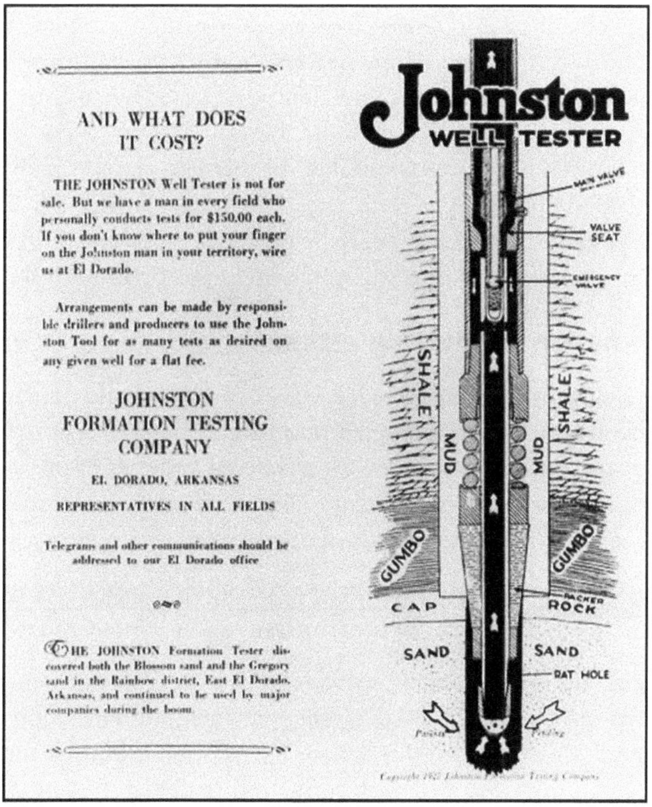

Johnston formation tester advertisement from 1927.

Formation Evaluation with Pre-Digital Well Logs
DOI: https://doi.org/10.1016/B978-0-12-820232-6.00017-0

17.1 Formation testing perspective

Today wireline formation testing tools provide versatility in matters of multiple packer assemblies and their ability to sample and pump fluids through, or out of, or into a virtual on-board laboratory. The tools available are packed with highly sensitive pressure gauges and allow for real time monitoring of basic fluid properties such as flowing fluid density, resistivity, and pH. Other sensors add a whole new level of sophistication with nuclear magnetic resonance, fluorescence, and optical spectroscopy that permits the scrutiny of flow samples. As a result today's analyst has the luxury of complete information on formation fluid content, formation water resistivity, hydrocarbon type, directional permeabilities, proximity of wellbores to barriers, and so on. The actual recovery of fluid samples is almost an afterthought.

With this in mind the pre-digital tools would seem almost primitive. They had but a single packer, no way to test if the seal between the packer and the formation was good (until it was too late!) and very limited sample chambers for retrieval of formation fluid samples. Certainly there was no way to avoid sampling formation fluids that came from the "contaminated" invaded zone. Any fluids recovered were thus mixtures of oil, gas, formation water, and mud filtrate. Despite these limitations quite sophisticated analysis was possible. In this section, the reader will be introduced to simple analysis tools and charts that allowed the formation testers of yesteryear to make very reasonable estimates of production potential in terms of flow rate, water cut, and gas/oil ratio (GOR).

17.2 Formation production estimates

When large samples were recovered, it was possible to predict formation productivity by analysis of the recovered materials. A mini-separator (Fig. 17.1) was used at the surface to measure the recovered volumes of oil, water, and gas. The water recovered was a mixture of mud filtrate and formation water. The amount of formation water was calculated from the relation

$$\%\,\text{formation water} = \frac{\text{ppm recovered water} - \text{ppm filtrate}}{\text{ppm formation water} - \text{ppm filtrate}}$$

Where all the water resistivities were known it was possible to use the chart shown on Fig. 17.2 to deduce the percentage of formation water in the recovered sample. Note that where R_w was an unknown the spontaneous potential (SP), coupled with formation temperature (and hence K), permitted the entry to the X-axis of the chart. The Y-axis value of the R_{mf}/R_{rf} ratio allowed a plotted point on the body of the chart that would fall on a line of formation water %.

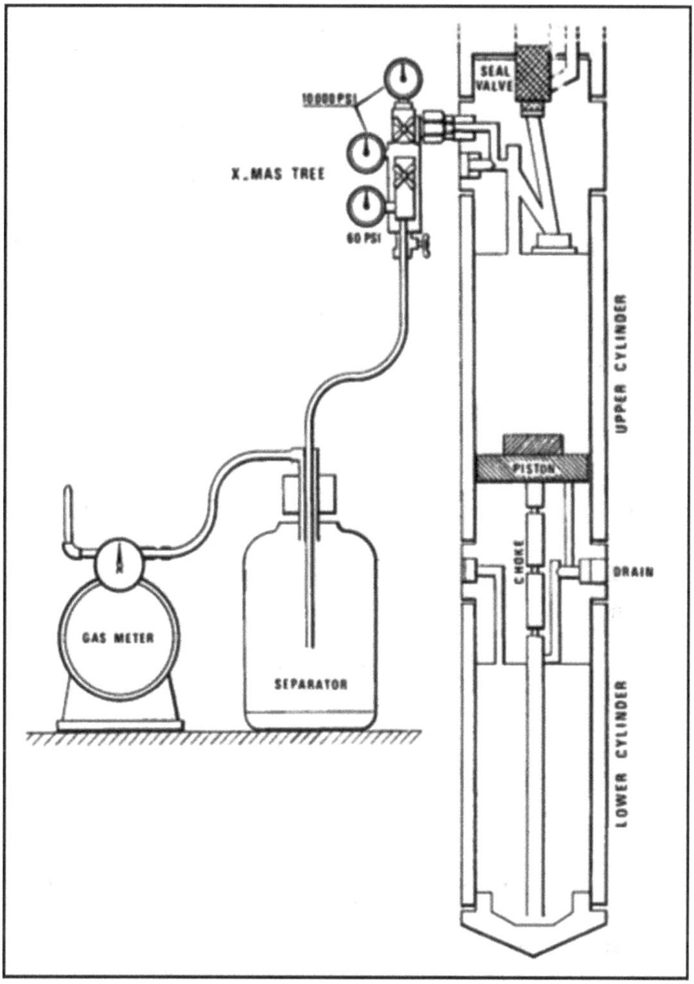

Figure 17.1 Formation tester sample recovery measurements. *Image courtesy Schlumberger.*

Empirical charts then linked recovered volumes to predicted production. Fig. 17.3 shows such a plot of oil recovery (in cc) versus gas recovery (in scf). Note that these empirical charts were built for specific sample chamber sizes. The one shown here was for a 2¾ gallon chamber. Note also that there are three areas delineated on the chart indicating formations that are gas, oil, and water productive.

An estimate of water cut can also be made using

$$\text{Water cut \%} = \frac{\text{Volume of formation water}}{\text{Volume of formation water} + \text{Volume of oil}} \times 100$$

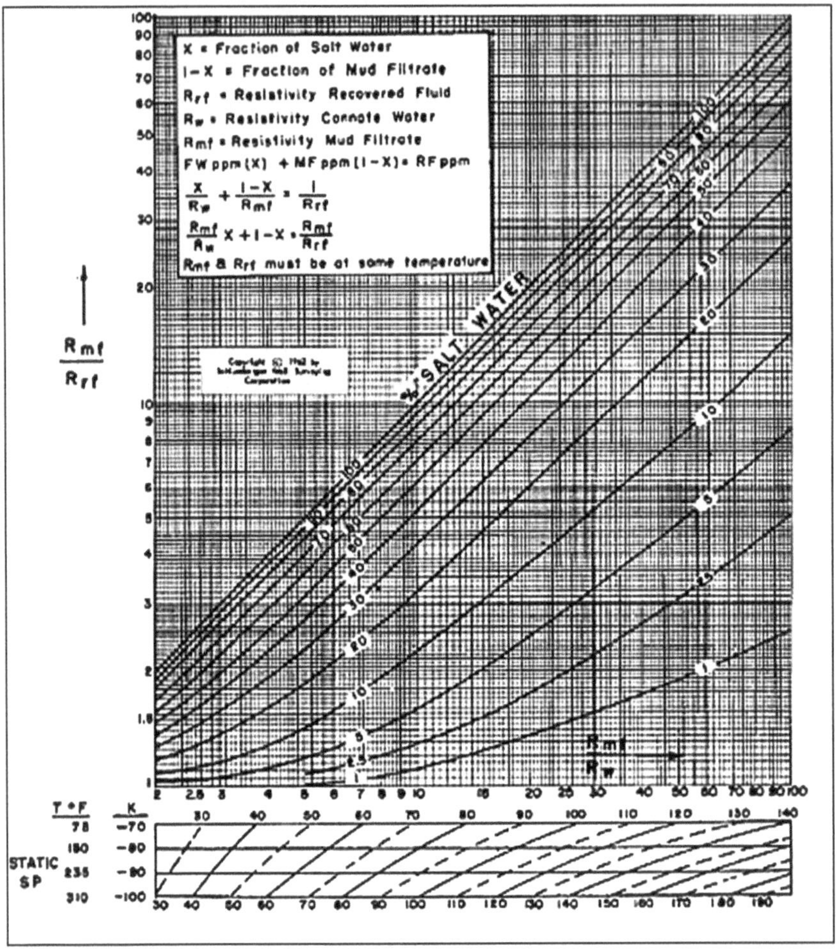

Figure 17.2 Determination of formation water percentage. *Image courtesy Schlumberger.*

17.3 Example FT recovery analysis

As an example a formation tester made the following recoveries as measured at the surface:

Gas: 23 scf

Oil: 7250 cc

Water: 1350 cc

Recovered fluid resistivity (R_{rf}) was 0.88 at 78°F.

Mud filtrate resistivity (R_{mf}) was 1.9 at 78°F.

Formation water resistivity (R_w) was 0.32 at 78°F.

We will use Fig. 17.2 to help predict water cut.

$R_{mf}/R_w = 5.94$ (see the green line on the chart)

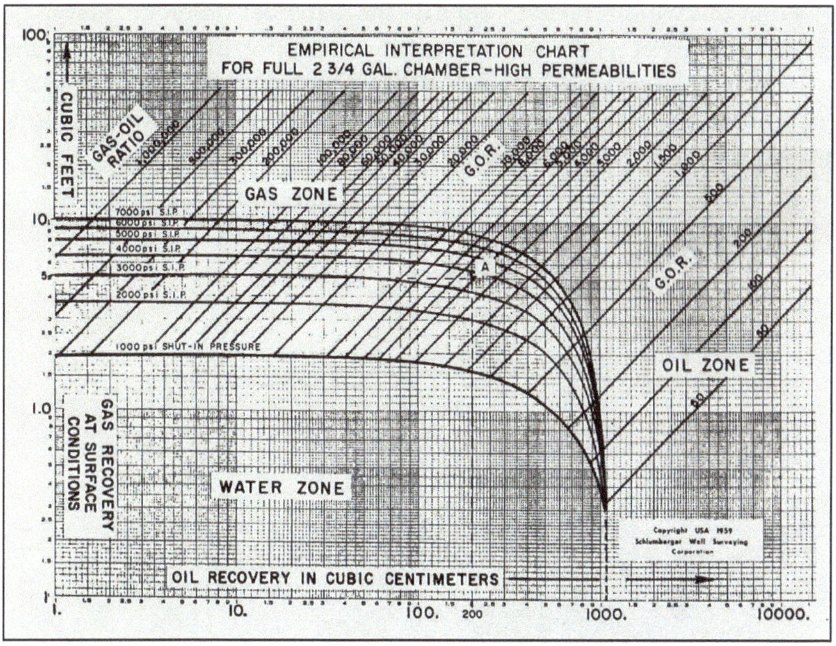

Figure 17.3 Empirical estimation of productivity. *Image courtesy Schlumberger.*

$R_{mf}/R_{rf} = 2.16$ (see the red line)

From which we read that the estimated formation water content of the recovered water is 21% (blue dashed line).

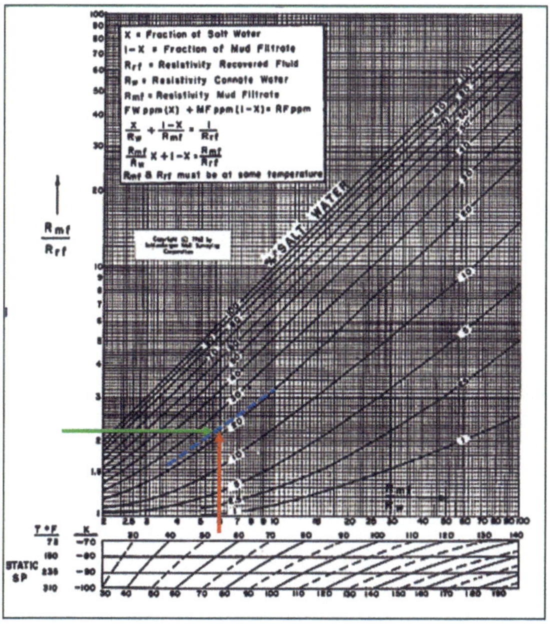

From this we can hope to estimate the water cut on production using the formula given in the text earlier which calls for the formation water volume and the sum of the water and oil volumes. The formation water volume will be the product of the total water recovered and the just estimated formation water content:

$$1350 \times 0.21 = 283.5 \text{ cc}$$

Hence:

$$\text{Water cut} = \frac{283.5}{1350 + 283.5} \times 100 = 3.76\%$$

Next, we will use Fig. 17.3 to predict oil and gas production.

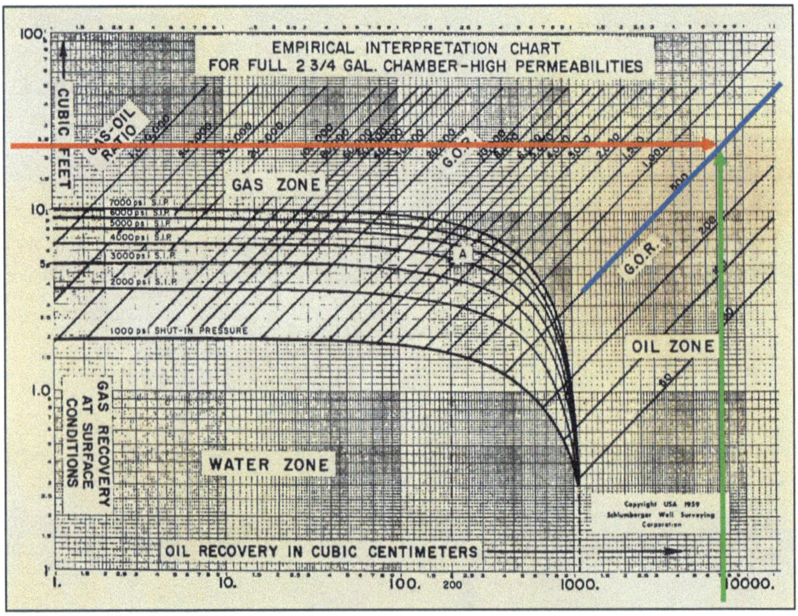

Recoveries of gas (red line) and oil (green line) plot in oil zone with GOR (blue line) = 500. In this case the well actually produced at 120 BOPD, no water, and GOR 700 scf/B.

Further reading

Dresser Atlas, 1980. Formation Multi-Tester Interpretation Manual. (Reference 06/80 9404).
Finklea, E.E., 1959. Formation Testing on Logging Cable - New Applications in Consolidated Formations. API Spring Meeting, Amarillo.
Lebourg, M., Fields, R.Q., Doh, C.A., 1956. A Method of Formation Testing on Logging Cable. TP No. 701-G, Fall Meeting. Soc. of Petr. Engineers of A.I.M.E., Los Angeles.
Schlumberger, 1966. Formation Tester Interpretation-Methods and Charts. Schlumberger.

CHAPTER 18

Early computer-processed interpretation

Abstract

Before computers made their way into wireline logging trucks, they at least made it into the head and field offices of wireline companies and those of their clients, the exploration and production companies. The calculation of porosity and water saturation was a straightforward algebraic manipulation that offered little challenge once the log data had been ingested into the computing machines of the time, at first by hand digitization onto 80-column IBM cards and then, later on, by machine digitization of log data curves from paper prints. In this chapter we will visit some of the ingenious ways that log users arrived at computer-processed log interpretations (CPIs).

An analog resistor network used to calculate electric log departure curves.
Image courtesy Schlumberger.

18.1 Historical perspective of computer-processed log interpretation

Three different groups were involved in computer-processed log interpretation (CPI). They were the oil and gas exploration and production (E&P) companies, the oilfield service companies and later, the third-party

Formation Evaluation with Pre-Digital Well Logs
DOI: https://doi.org/10.1016/B978-0-12-820232-6.00018-2

consulting and software provider companies. It could be said that CPI arrived *before* the start of the true age of digitally recording well logs themselves. Even though the field recordings were, for the most part, still analog, the processing of the data by computer came first. Highly intensive computations, such as dipmeter processing, required a "mainframe" computer which was, at that time, a huge investment and thus only a few "data centers" were equipped to perform such functions. Moreover, such processing could only be done by the same party that had recorded the data at the well site. The E&P companies were probably the earliest "drivers" of the art and science of digital log processing for petrophysical "answers." The pioneers worked at the E&P research departments that were active in the 1970s and 1980s before the industry switched to "outsourcing" and the closing down of their in-house projects altogether. The second group consisted of the wireline service companies. Log computing centers were opened to process logs that were recorded indirectly on punched paper tape or on ½″ magnetic tape. The analog prints were still the primary log records, but the service companies recorded some of the raw data using proprietary data formats. They also sought competitive advantages by use of their in-house research centers to produce log analysis software routines that could feed the growing appetite from the E&P companies for "answer products" that they were willing to pay for. The third-party services were eventually populated from the ranks of E&P and wireline service company employees, "downsized" on the recommendations of business management companies manned by eager possessors of MBA degrees from prestigious business schools.

18.2 Under-, exactly-, and over-determined systems

In the matter of log interpretation, it is common to divide the mathematical problem into three categories:

- Under-determined systems,
- Exactly determined systems, and
- Over-determined systems.

Under-determined systems refer to those cases where the analyst finds that there are more unknowns than there are independent logging measurements. For example, the unknowns might be porosity, water saturation, rock type, hydrocarbon type, water salinity, and shale content. A total of six. If the only logging measurements available are a gamma ray, spontaneous potential (SP), and a resistivity curve, then an exact solution is not possible without making assumptions about the geology and the fluids in the pore space. Under such circumstances, the log analyst looks for low gamma ray coupled with high resistivity readings—and hopes for the "qualitative" best answer. For this neither a computer nor a CPI is needed or called for.

Exactly determined systems refer to those cases where the analyst is blessed with the same number of knowns as unknowns. An example might be where the logs are run in a hard rock environment where the matrix is known to contain limestone, dolomite, sandstone, and anhydrite, the water salinity is well known, and the hydrocarbon type is oil. The unknowns are the proportions of different matrix components, the porosity, and the water saturation—all together six items. If the well has been logged with density, neutron, sonic, gamma ray, and resistivity logs an exact solution is possible since there are five measurements plus the knowledge that the sum of the volume fractions of the subcomponents of the mixture must be 1. The earliest CPI routines used a simple matrix inversion technique that will be described in more detail below (see the Chavaroo method).

Over-determined systems refer to those occasions when there are more measurements than there are unknowns. Such cases became the norm, rather than the exception, as modern logging technology kept adding new parameters to the logger's arsenal. For example, the addition of a Pe curve or $\Delta t_{\mathrm{Shear}}$ or the breakout of the GR into components K, U, and Th. Although this might be considered a blessing, it merely highlighted the fact that logging measurements are inexact and that the response of any given tool to a mineral or fluid is not well known sometimes or even subject to measurement error due to environmental conditions. As a result, many similar, but different, results may be obtained from the same raw logging data set and the art of the analyst is then to determine which of the many petrophysical solutions is the *most probable* one. The CPI programmers soon mastered these over-determined cases with sophisticated error minimization routines under such names as Multimin, Global, Ultra, and others that will be discussed in detail later.

Falling into none of the abovementioned categories are a separate set of CPI routines and software products that used complex logic complete with "if, then, else" branches that attempted to make the most intelligent choices for matrix components, water saturation, permeability, and so on in "messy" situations such as shaly sands and/or fractured carbonates. These too will be discussed in detail later (see Saraband, Coriband, Dual Water, etc.).

18.3 Matrix inversion and the Chavaroo method (old wine in new bottles)

Fig. 18.1 illustrates the output of the early "matrix inversion" programs used in the Permian Basin in the late 1960s. On the left-hand side of the presentation is a lithology track which shows the relative proportions (in this case) of sandstone (SIL), gypsum (GYP), anhydrite (ANH), dolomite

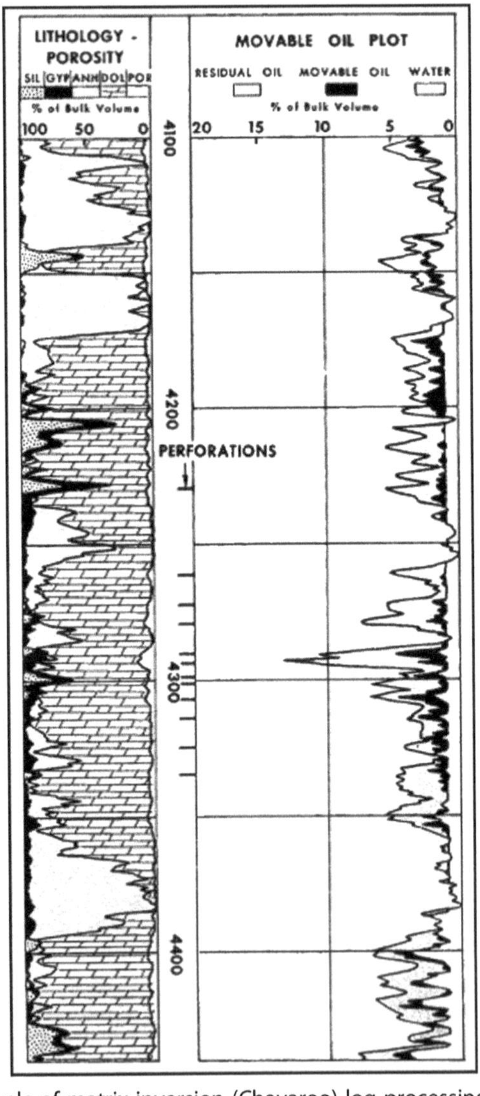

Figure 18.1 Example of matrix inversion (Chavaroo) log processing. *Reprinted by permission of the SPE-AIME from Burke et al., 1966.*

(DOL), and porosity (POR). The two right-hand tracks show the breakdown of the pore space into water, residual oil, and movable oil. It is noticeable that the section is predominantly low porosity dolomite (~5%) with some anhydrite beds.

The logic behind this kind of processing is based on the solution of a series of simultaneous equations relating the log measurements of bulk density (ρ_B), interval travel time (Δt), and neutron porosity (ϕ_N), and so on to the volume fractions of limestone (V_L), dolomite (V_D), sandstone (V_S), and porosity (ϕ)—for example.

$$\rho_b = V_L \times 2.71 + V_D \times 2.87 + V_S \times 2.66 + \phi \times 1.0$$

$$\Delta t = V_L \times 47.5 + V_D \times 43.5 + V_S \times 55.5 + \phi \times 189$$

$$\phi_N = V_L \times 0 + V_D \times 0.07 + V_S \times (-0.05) + \phi \times 1$$

$$1 = V_L + V_D + V_S + \phi$$

The modern log analyst has the advantage today with a feature of Microsoft Excel that painlessly allows the inversion of a matrix as shown in Fig. 18.2.

Matrix A is set up with the endpoints for each logging tool sensor and each volume fraction that is to be determined. In the example shown there are three log measurements and three possible matrix materials plus porosity. A second matrix (B) is set up with the actual log measurements. Matrix C is the inverse matrix of A which is simply accomplished using the command "MINVERSE(A)." The final step is to set up the "answer" Matrix D using the Excel command "MMULT(C,B). The reader is encouraged to attempt to duplicate the matrix manipulations documented

Matrix A				
	Lime	Dol	Sand	Poro
ρB	2.71	2.87	2.66	1
Δt	47.5	43.5	55.5	189
ϕN	0	0.07	−0.05	1
	1	1	1	1

B
Logs
2.49
69
0.165
1

Matrix C			
Inverse of A			
−13	−0	3.9	44
6.7	0.1	2.6	−21
6.1	0.1	−7	−23
−0	0	0.5	0.3
MINVERSE (A)			

D	
Answers	
0.40	Lime
0.30	Dol
0.15	Sand
0.15	Poro
MMULT (C,B)	

Figure 18.2 Use of modern Excel for log matrix inversions.

Matrix A				
	Sand	Shale	Oil	Water
ρB	2.65	2.5	0.8	1
GR	20	120	0	0
ϕN	0	0.45	0.8	1
	1	1	1	1

B
Logs
2.3
40
0.25
1

Matrix C			
Inverse of A			
0.4	−0	−0	0
−0	0	0.1	0
−2	0	−3	5
1.6	−0	3.4	−4
MINVERSE (A)			

D	
Answers	
0.59	Sand
0.23	Shale
0.14	Oil
0.03	Water
MMULT (C,B)	

Figure 18.3 Matrix inversion for shaly sand.

here and observe the effects of minor changes in the log response "endpoints" and the effects on the "answers" of minor changes in the log readings themselves, as entered in Matrix B.

The example, shown in Fig. 18.2, was for complex lithology, but the method works equally well for shaly sand sequences. Fig. 18.3 shows the setup for finding porosity in a shaly sand given measurements of density, gamma ray, and neutron. As can be seen, the method is highly flexible and lends itself to formation evaluation from pre-digital logs. All the analyst has to do, after the initial setup of the matrices, is read log values from the paper prints and type them into the Excel spreadsheet. This is, indeed, old wine in a new bottle.

18.4 E&P company research CPI

The major E&P companies were not slow to perfect their own in-house log processing programs. Typical of these was the Herb Lindley program which came out of Amoco's Tulsa Research group. The program was based on a series of commands keyed onto 80-column IBM cards. Each card would specify the start and end depths and the function to be performed. This might be to compute density porosity from the RhoB curve. In specific columns of the input IBM cards the user would have to specify rho-matrix, rho-fluid, and so on. The completed deck of IBM cards would then be delivered to a "high priest" behind a glass window at the computing center. If the job control language (JCL) was correctly parsed then the deck would be fed into a "main frame" and, with luck, a completed log analysis could be expected in about 24 hours. The original Herb Lindley program was later taken under the wing of Jay Patchett and released again as *Amolog*.

For quick wellsite analysis, other methods were used that incorporated a digitizing tablet that could input the logs directly to a mini-computer on-site. The Amoco version for this type of log analysis program was called *Inlan*. It was more user-friendly and flexible than the mainframe software and allowed the user to add or modify the log interpretation logic to fit the actual case in hand. Other E&P companies developed and used similar CPI tools.

18.5 Wireline service company CPI products

In parallel with the E&P company log processing efforts the wireline service companies invested great effort to provide "answer products" that would, at the same time, offer valuable interpretation of the logs but also a source of income from computing fees, charged at regional log computing centers, that quickly sprang up in the late 1970s and early 1980s.

Under the Schlumberger banner the most widely known and used were *Saraband* (for shaly sands) and *Coriband* (for complex lithology). The programs could accept digitized version of the optical logs or digital versions as the wellsite taping of logs became more prevalent. The logic used in these "flag ship" products was quite complex and, although described in published papers, still left some of the more subtle petrophysical nuances hidden. This gave rise to accusations from the E&P companies that the wireline companies were selling them a "black box."

Fig. 18.4 shows the output from a Saraband computation. The presentation is on four tracks. Track 1 shows the shale fraction, V_{sh} along with an estimate of permeability. Note that while the shale content increases to the right, the permeability increases to the left. Track 2 provides a "hydrocarbon analysis" which includes the water saturation (S_w), the hydrocarbon volume $\{\phi(1 - S_w)\}$, and the hydrocarbon weight (hydrocarbon volume multiplied by ρ_{hy}).

This presentation facilitates the picking of both the oil/water contact and a gas/oil contact, if such exists. On this present example both are indicated. Track 3 is a porosity analysis that includes both a porosity curve and a water-filled porosity curve (ϕS_w). The area between the two curves is shaded black in the example. Track 4 is a bulk volume analysis that divides the formation components into its constituent fractions of clay, sand, and pore space. Based on this analysis the upper logged section is a relatively clean sandstone with 30% porosity and an irreducible water saturation of around 10%.

Fig. 18.5 shows an example of Schlumberger's Coriband. Track 1 offers insight into the formation characteristics by plotting the average grain density along with indications of any secondary (fracture) porosity.

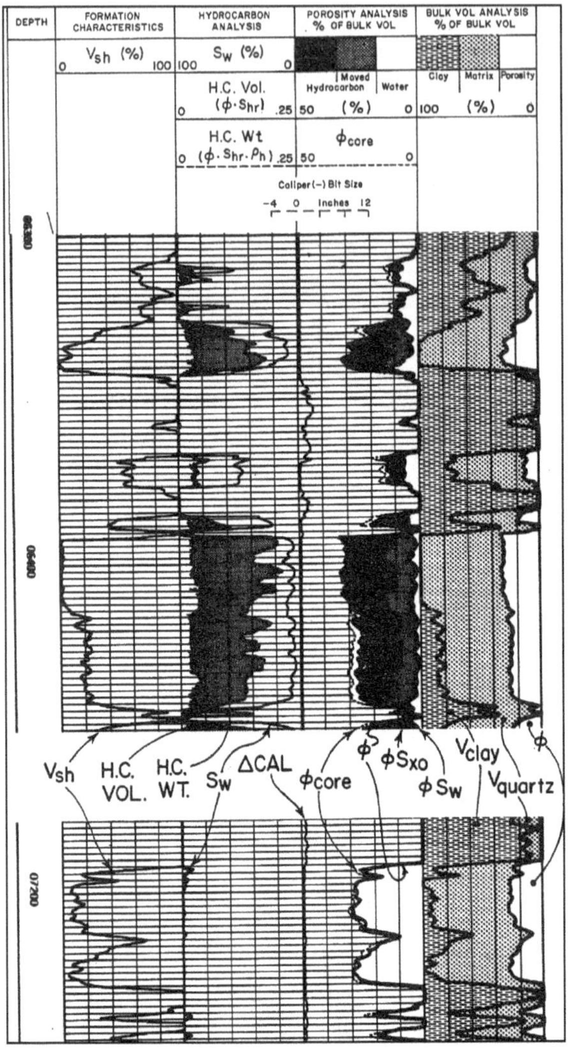

Figure 18.4 Output from a Saraband. ***(Shaly Sand CPI)****—Image courtesy Schlumberger.*

Note that in this example the predominant formation grain density as close to 2.8 g/cc, indicating dolomitic limestone. Track 2 shows the water saturation (S_w). The lower part of the logged section is hydrocarbon bearing and fractured or vuggy. Track 3 gives a porosity analysis with residual and moved hydrocarbon as well as connate water. It would appear in this example that most of the hydrocarbon is residual. Track 4 gives the formation analysis by volume and maps three components, clay, porosity, and undesignated matrix. It is left to the user to refer to the first track and decide what the rock matrix is, based on the grain density plotted there.

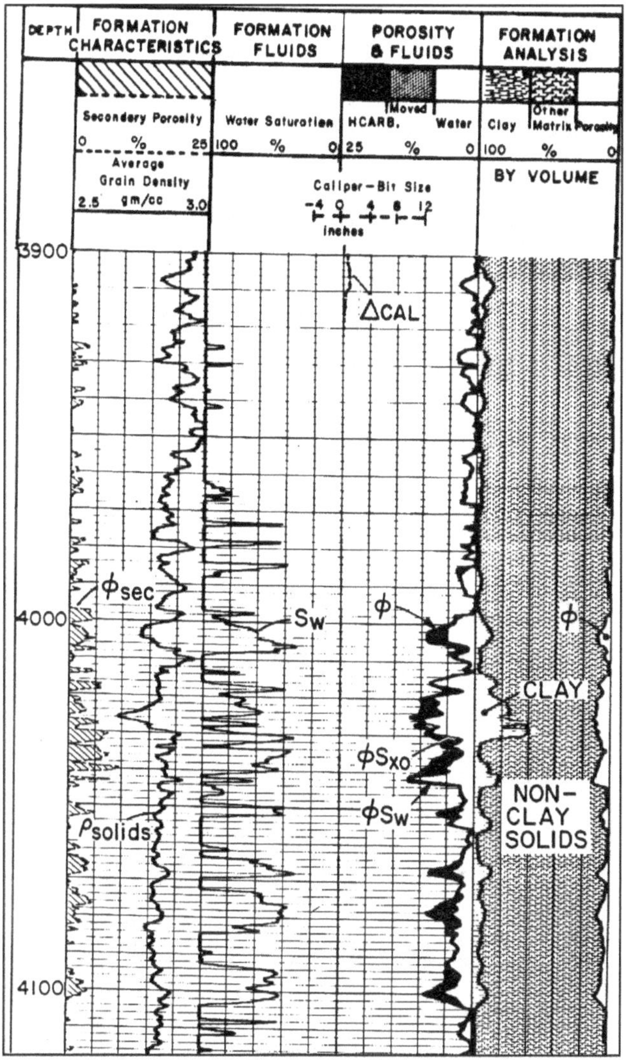

Figure 18.5 Coriband presentation for complex lithology CPI. *Image courtesy Schlumberger.*

18.6 Wellsite CPI products

The current sections would not be complete without mention of the wellsite CPI products that appeared soon after the end of the pre-digital era with the introduction of logging truck borne computers that could not only control and record the logging parameters digitally but could also produce computed petrophysical "answers" at the wellsite itself. These were log interpretation routines adapted from the packages that were available at the fixed site computing centers to run on the smaller computers that were carried on the logging trucks. Fig. 18.6 gives an

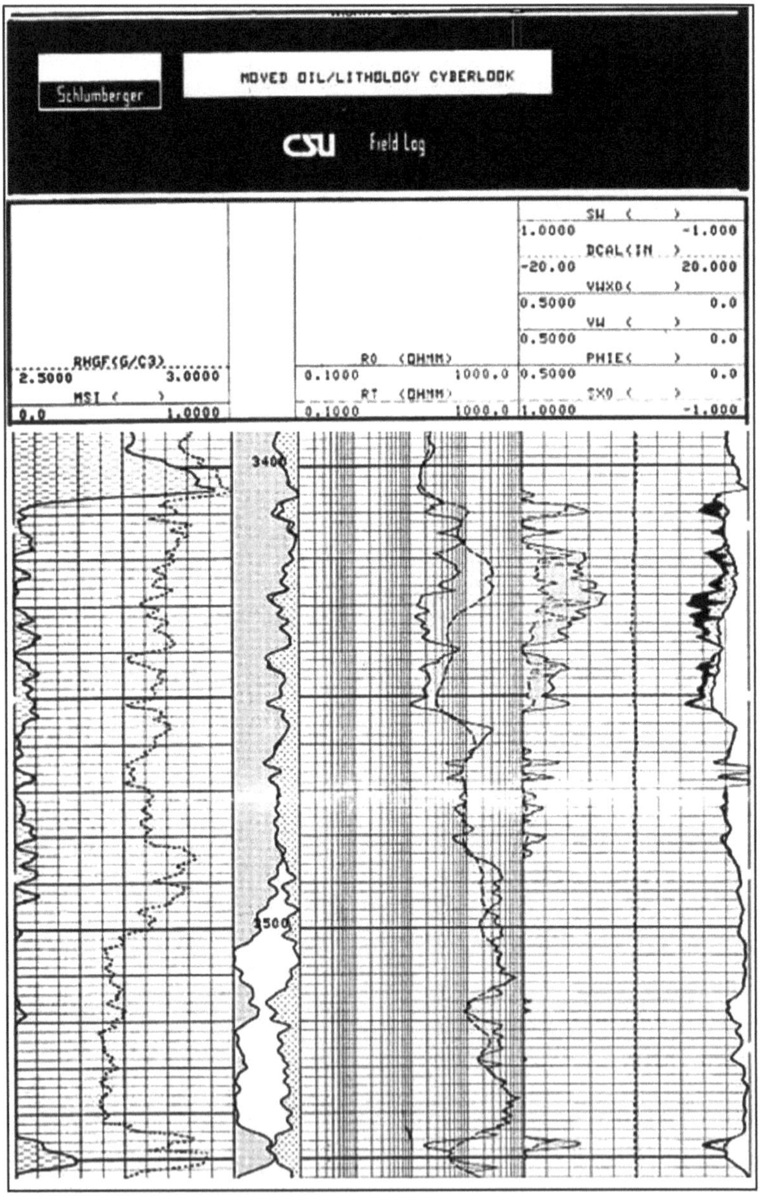

Figure 18.6 Wellsite CPI offered under the name cyberlook. *Image courtesy Schlumberger.*

example of one of these which went by the name of *Cyberlook*. The basis for the Cyberlook analysis was the "dual water" system which considered that the salinity of the "free" water in the pore space was different from the "bound" water associated with the clays present in shaly sands. The Archie water saturation equation was thus modified to accommodate two water resistivities, R_{wf} and R_{wb}.

On the Cyberlook output plot, shown in Fig. 18.6, the left-hand track shows the lithological indicators of grain density and "shale index" (V_{sh}). The central track displays R_t and R_o and shades in the areas between them in hydrocarbon bearing zones where they diverge. The two right-hand half-tracks show S_w, a Δ caliper (hole-bit size} and a breakdown between water and hydrocarbon-filled pore space.

Wellsite presentations of this ilk were more than sufficient for decision-making on such questions as where to run formation tests, collect sidewall cores, or even whether or not to run and cement casing. If the printouts from any of these "computer-processed interpretations" have survived and are in the filing cabinets that the reader is charged with "mining," then a few words of caution are in order. The computer is superb when it comes to making complex calculations repetitively as, for example, every 6 in. over a logged interval of 5000 ft. However, if the parameters chosen for the key interpretation are incorrect the outputs will likewise be wrong. So, it is a case of caveat emptor. In the light of what is known today about that old field you are charged with analyzing do be sure that at least the R_w used, for example, is correct!

18.7 Sliderules and programmable calculators

Log analysts of the pre-digital era used analog methods (before the introduction of any of the digital log elaboration methods mentioned in this chapter) based on the *sliderule*. Few of the readers of this work will ever have had to use a sliderule. Thus, a reminder that adding the logarithm of two numbers will give the logarithm of their product and the difference between them the result of dividing one by the other. Since most of the mathematics of log analysis is based on multiplication and division it was only natural that log analysis sliderules would come to be a "standard issue" in the briefcases of wellsite geologists and loggers. Fig. 18.7 shows just such a sliderule. The side shown includes means to calculate R_w from the SP, input porosity, and the "*m*" exponent. The reverse side allows for the input of R_t and the reading of S_w.

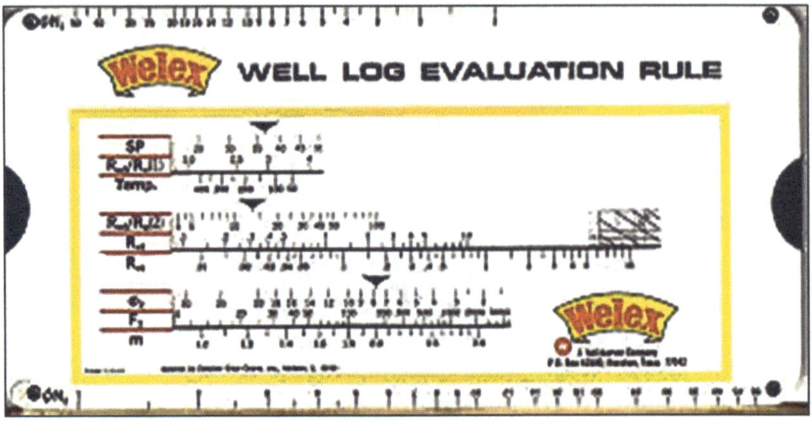

Figure 18.7 Log analysis sliderule from the 1950s and 1960s.

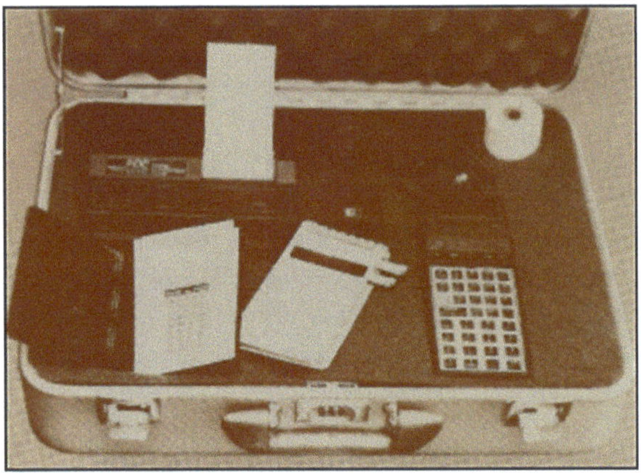

Figure 18.8 Programmable calculator used for log analysis.

Toward the end of the pre-digital era, handheld calculators became readily available and the more advanced versions allowed for programming of simple log analysis routines. Fig. 18.8 shows one such popular model in use at that time, based on the HP 41C.

References

Burke, J.A., Curtis, M.R., Cox, J.T., 1966. Computer processing of log data enables better production in chaveroo field. SPE Paper 1576, Presented at the 41st Annual Meeting. 2–5 October, Dallas.

Clavier, C., Coates, G., Dumanoir, J., 1977. Theoretical and experimental bases for the dual water model for the interpretation of shaly sands. 52nd Annual Fall Technical

Conference and Exhibition. Paper 6859, October. Society of Petroleum Engineers of AIME, Denver, Colorado.
Poupon, A., Clavier, C., Dumanoir, J., Gaymard, R., Misk, A., 1970. Log analysis of sand shale sequences - a systematic approach. J. Pet. Tech. 249, 867–881.
Poupon, A., Hoyle, W.R., Schmidt, A.W., 1971. Log analysis in formations with complex lithologies. J. Pet. Tech. 23 (8), 995–1005.
Raymer, L.L., Biggs, W.P., 1963. Matrix characteristics defined by porosity computations. SPWLA 4th Annual Logging Symposium. May.
Savre, W.C., 1963. Determination of a more accurate porosity and mineral composition in complex lithologies with the use of sonic, neutron and density surveys. J. Pet. Tech. 16, 945–959.
Schmidt, A.W., Land, A.G., Yunker, J.D., Kilgore, E.C., 1970. Applications of the coriband technique to complex lithologies. JPT 867.

Bibliography

Technical Pre-digital References to be included in the End Matter

Anon, 1936. History of geophysics. Pet. Times, London 36 (916), 1.

Guyod, H., 1955. Electrical analog of resistivity logging. Geophysics 20 (3), 615.

Hamilton, R.G., 1960. The revolution in well logging. Oil Gas J. 58 (26), 187.

Historical Log Interpretation Charts, 1947. The Denver Well Logging Society – A Collection of 7 CDs with Charts from Schlumberger. PGAC, Dresser Atlas, Western Atlas, and Atlas Wireline Services, Halliburton, Welex, GO, Gearhart, Elgen.

Landes, K.K., 1951. Petroleum Geology. John Wiley and Sons, New York, p. 69.

1957Log Interpretation Charts. Schlumberger Well Surveying Corporation, Houston.

Panyity, L.S., 1937. New method for taking core samples at well. Oil Gas J. 36 (16), 46.

Passey, Q., Creaney, S., Kulla, J., Moretti, F., Stroud, J., 1990. A practical model for organic richness from porosity and resistivity logs. AAPG Bulletin 74 (12), 1777–1794.

Rust Jr., W.M., 1938. A historical review of electrical prospecting methods. Geophysics 3 (1), 2.

Schlumberger, 2005: Historical Log Interpretation Charts, 2005 ed. (05-FE-006).

Tittle, C.W., 1961. Theory of neutron logging. Geophysics 26, 27–29.

Tixier, M.P., Martin, M., 1954. History of logging and future development. Can. Oil Gas Ind. 7 (8).

Zhao, H., Givens, N.B., Curtis, B., 2007. Thermal maturity of the Barnett Shale determined from well-log analysis. AAPG Bulletin 91 (4), 535–549.

Non-Technical Background Reading List to be included in the End Matter

Allaud, L.A., Martin, M.H., 1977. Schlumberger, The History of a Technique. John Wiley & Sons.

Boatright, M.C., 1963. Folklore of the Oil Industry. Southern Methodist University Press, Dallas.

Cunningham Craig, E.H., 1912. Oil-Finding – An Introduction to the Geological Study of Petroleum. Longmans, Green and Co. New York.

Eavenson, H.N., 1935. Coal Through The Ages. American Institute if Mining and Metallurgical Engineers (Incorporated), The Maple Press Company, York, PA.

Geikie, A., 1909. Charles Darwin as Geologist – The Rede Lecture. Cambridge University Press, London.

Hatley, A.G. (Ed.), 1995. The Oil Finders - A Collection of Stories about Exploration. Centrex Press, Utopia, Texas.

Houston Chapter, 1979. The Art of Ancient Log Analysis. Society of Professional Well Log Analysts.

Hurt III, H., 1982. Texas Rich – The Hunt Dynasty from the early Oil Days through the Silver Crash. W. W. Norton & Company, New York.

Jost, R., 2006. September 5th, 1927: A world premiere. Petrophysics 51 (4), 1–3.

Lenzner, R., 1986. The Great Getty – The Life and Loves of J. Paul Getty Richest Man in the World. Crown Publishers, Inc, New York.

Mathews, J.J., 1951. Life and Death of an Oilman -The Career of E.W. Marland. University of Oklahoma Press, Norman, OK.

Mau, M., Edmundson, H., 2015. Groundbreakers: The Story of Oilfield Technology and the People Who Made it Happen. Fast-Print Publishing of Peterborough, England.

McKee, R.H., 1925. Shale Oil. American Chemical Society, The Chemical Catalog Company, Inc., New York.
Olien, R.M., Hinton, D.D., 2009. Wildcatters – Texas Independent Oilmen. Texas A&M University Press.
Oppenheim, V., 1958. Explorations East of the High Andes – (From Patagonia to the Amazon). Pageant Press Inc., New York.
Oristaglio, M., Dorozynski, A., 2009. A Sixth Sense –The Life and Science of Henri-Georges Doll. Overlook Duckworth, New York.
Presley, J., 1978. A Saga of Wealth – The Rise of the Texas Oilmen. G.P. Putnam's Sons, New York.
Schlumberger, A.G., 1982. The Schlumberger Adventure. Arco Publishing, Inc, New York.
Schlumberger, 2007. 80 Years of Innovation. Schlumberger.
Sheldon Knowles, R., 1978. The Greatest Gamblers – The Epic of American Oil Exploration. University of Oklahoma Press, Norman.
Tait Jr., S.W., 1946. The Wildcatters – An Informal History of Oil-Hunting in America. Princeton University Press.
The Pennsylvania State College, 1933. Petroleum and Natural Gas – Production and Natural gas Series. State College, Pennsylvania.
Williamson, J.W., 1927. In a Persian Oil Field - A Study in Scientific and Industrial Development. Ernest Benn Limited, London.

Index

Note: Page numbers followed by "*f*" and "*t*" refer to figures and tables, respectively.

T

U

W

Z

FSC
www.fsc.org
MIX
Papier | Fördert
gute Waldnutzung
FSC® C083411

Kolibri 360 GmbH
Ferdinand-Jühlke-Straße 7
99095 Erfurt, Deutschland
produktsicherheit@kolibri360.de